Time
A Traveler's Guide

Works by Clifford A. Pickover

The Alien IQ Test
Black Holes, A Traveler's Guide
Can You Escape?
Chaos in Wonderland: Visual Adventures in a Fractal World
Computers, Fractals, Chaos (Japanese)
Computers, Pattern, Chaos, and Beauty
Computers and the Imagination
Future Health: Computers and Medicine in the 21st Century
Fractal Horizons: The Future Use of Fractals
Frontiers of Scientific Visualization (with Stu Tewksbury)
Keys to Infinity
The Loom of God
Mazes for the Mind: Computers and the Unexpected
Mit den Augen des Computers
The Pattern Book: Fractals, Art, and Nature
Spider Legs (with Piers Anthony)
Spiral Symmetry (with Istvan Hargittai)
Strange Brains and Genius
Visions of the Future: Art, Technology, and Computing in the 21st Century
Visualizing Biological Information

Time

A Traveler's Guide

Clifford A. Pickover

OXFORD
UNIVERSITY PRESS

OXFORD
UNIVERSITY PRESS

Oxford New York
Athens Auckland Bangkok Bogotá Buenos Aires Calcutta
Cape Town Chennai Dar es Salaam Delhi Florence Hong Kong Istanbul
Karachi Kuala Lumpur Madrid Melbourne Mexico City Mumbai
Nairobi Paris São Paulo Singapore Taipei Tokyo Toronto Warsaw

and associated companies in
Berlin Ibadan

First published by Oxford University Press, Inc., 1998

First issued as an Oxford University Press paperback, 1999
198 Madison Avenue, New York, New York 10016

Library of Congress Cataloging-in-Publication Data
Pickover, Clifford A.
Time: a traveler's guide / by Clifford A. Pickover.
p. cm. Includes bibliographical references
ISBN 0-19-512042-6
ISBN 0-19-513096-0 (pbk.)
1. Space and time. 2. Time.
3. Time Travel. I. Title.
QC173.59.S65P53 1998 530.11—dc21 97-22369

Design by Nanker & Phelge

1 3 5 7 9 10 8 6 4 2

Printed in the United States of America
on acid-free paper

This book is dedicated to time travelers from the future
who are reading this book in the 20th Century.
The curtain of the Universe is rent and shattered,
The splendor-winged worlds disperse like wild doves scattered.
Percy B. Shelley
Hellas, A Lyrical Drama, 1821

Acknowledgments

> Imagine the outcry about the waste of tax-payers' money if it were known that the National Science Foundation were supporting research on time travel. For this reason, scientists working in this field have to disguise their real interest by using technical terms like "closed time-like curves" that are code for time travel.
>
> —Stephen Hawking

> How little do you mortals understand time. Must you be so linear, Jean-Luc?
>
> —Q to Picard, in *Star Trek*'s "All Good Things"

> This present universe has evolved from an unspeakably unfamiliar earlier condition, and faces a future extinction of endless cold or intolerable heat. The more the universe seems comprehensible, the more it also seems pointless. . . . The effort to understand the universe is one of the few things that lifts human life a little above the level of farce, and gives it some of the grace of tragedy.
>
> —Steven Weinberg

Acknowledgments

I owe a special debt of gratitude to Professor Paul J. Nahin, time-travel expert, for his wonderful past books and papers from which I have drawn many facts and formulas about time travel. I heartily recommend his book *Time Machines* for further information on time travel in physics, metaphysics, and science fiction. I also thank Clay Fried and Robert Stong for their advice and comments.

The fragments of Chopin musical scores come from *Frederic Chopin Nocturnes and Polonaises* (The Paderewski Edition) edited by Ignacy Jan Paderewski, Ludwik Bronarski, and Jozef Turczynski, published by Dover Publications, and used with permission.

Contents

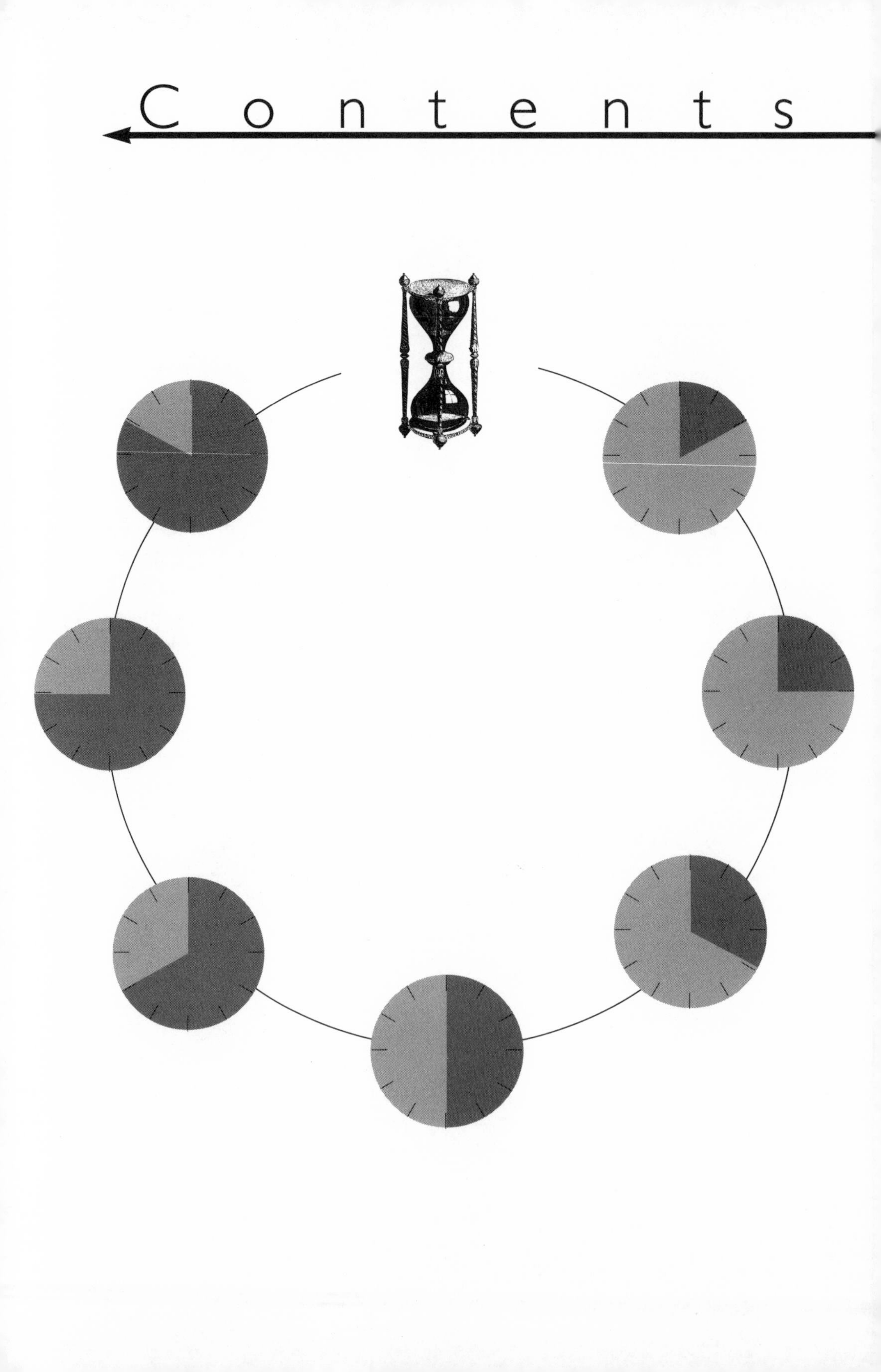

Chopin's piano was to him what a pony is to a cowboy.
Chopin had grown to be part of it. He knew exactly what
he could do with it, how much it would stand, how far and
how fast he could drive it without forcing it to collapse.
Having mastered it until it would obey his slightest whim,
he used it — consciously or unconsciously — for but one
single purpose — to give expression to that love which
bound him and his fellow exiles to their native soil.
—Hendrik Willem Van Loon, *The Arts,* 1937

We still cannot say what time is; we cannot agree whether there
is one time or many times, cannot even agree whether time is
an essential ingredient of the universe or whether it is
the grand illusion of the human intellect.
—Davis and Hersh, *Descartes' Dream,* 1986

There was a young lady named Bright
Who traveled much faster than light.
She started one day
In the relative way,
And returned on the previous night.
—A. H. Reginald Buller

Preface

> The job of science is to enable the inquiring mind to feel at home in a mysterious universe.
>
> —Lewis Carroll Epstein, *Relativity Visualized*

> Time is a relationship that we have with the rest of the universe; or more accurately, we are one of the clocks, measuring one kind of time. Animals and aliens may measure it differently. We may even be able to change our way of marking time one day, and open up new realms of experience, in which a day today will be a million years.
>
> —George Zebrowski, *OMNI,* 1994

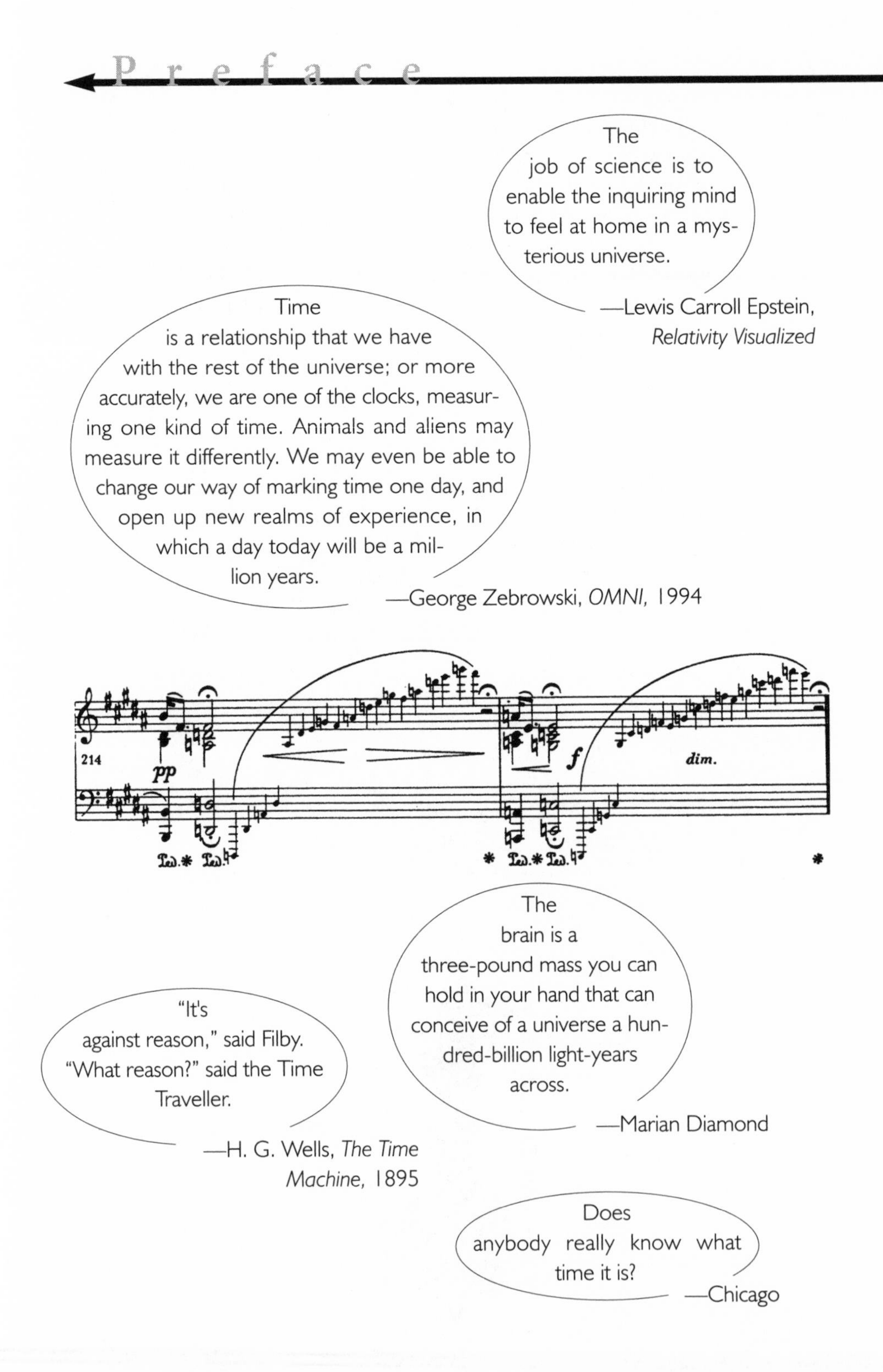

> The brain is a three-pound mass you can hold in your hand that can conceive of a universe a hundred-billion light-years across.
>
> —Marian Diamond

> "It's against reason," said Filby. "What reason?" said the Time Traveller.
>
> —H. G. Wells, *The Time Machine,* 1895

> Does anybody really know what time it is?
>
> —Chicago

Preface

The Quest for Eternity

What is time? Is time travel possible? For centuries, these questions have intrigued mystics, philosophers, and scientists. Much of ancient Greek philosophy was concerned with understanding the concept of eternity, and the subject of time is central to all the worlds' religions and cultures. Can the flow of time be stopped? Certainly some mystics thought so. Angelus Silesius, a sixth-century philosopher and poet, thought the flow of time could be suspended by mental powers:

> Time is of your own making;
> its clock ticks in your head.
> The moment you stop thought
> time too stops dead.

This book is mostly about the *science* of time travel and touches only briefly on mysticism. However, the line between science and mysticism sometimes grows thin. Today, physicists would agree that time is one of the strangest properties of our universe. In fact, there is a story circulating among scientists of an immigrant to America who has lost his watch. He walks up to a man on a New York street and asks, "Please, Sir, what is time?" The scientist replies, "I'm sorry, you'll have to ask a philosopher. I'm just a physicist."

Most cultures have a grammar with past and future tenses, and also demarcations like seconds and minutes, and yesterday and tomorrow. Yet we cannot say exactly what time is. Although the study of time became scientific during the time of Galileo and Newton, a comprehensive explanation was given only in this century by Einstein, who declared, in effect, time is simply what a clock reads. The clock can be the rotation of a planet, sand falling in an hour glass, a heartbeat, or vibrations of a cesium atom. A typical grandfather clock follows the simple Newtonian law that states that the velocity of a body not subject to external forces remains constant. This means that clock hands travel equal distances in equal times. While this kind of clock is useful for everyday life, modern science finds that time can be warped in various ways, like clay in the hands of a cosmic sculptor.

Science-fiction authors have had various uses for time machines including: dinosaur hunting, tourism, visits to one's ancestors, and animal collecting. Ever since the time of H. G. Wells's famous novel *The Time Machine* (1895), people have grown increasingly intrigued by the idea of traveling through time. (I was lucky enough to have chats with H. G. Wells's grandson, who told me that his grandfather's book has never been out of print, which is rare for a book a century old.) In the book, the protagonist uses a "black and polished brass" time machine to gain mechanical control over time as well as return to the present to bring back his story and assess the consequences of the present on the future. Wells was a graduate of the Imperial College of Science and Technology, and scientific language permeates his discussions. Many believe Wells's book to be the first story about a time machine, but seven years before twenty-two-year-old Wells wrote the first version of *The Time Machine,* Edward Page Mitchell, an editor of the New York *Sun,* published "The Clock That Went Backward." One of the earliest methods for fictional time travel didn't involve a machine; the main character in Washington Irving's "Rip Van Winkle" (1819) simply fell asleep for decades. King Arthur's daughter Gweneth slept for 500 years under Merlin's spell.

Ancient legends of time distortion are, in fact, quite common. One of the most poetic descriptions of time travel occurs in a popular medieval legend describing a monk entranced for a minute by the song of a magical bird.

Figure 1 Sir Isaac Newton. For Newton, both space and time were absolute. Space was a fixed, infinite, unmoving metric against which absolute motions could be measured. Einstein changed all this with his relativity theories, and once wrote, "Newton, forgive me."

When the bird stops singing, the monk discovers that several hundred years have passed. Another example is the Moslem legend of Muhammad carried by a mare into heaven. After a long visit, the prophet returns to Earth just in time to catch a jar of water the horse had kicked over before starting its ascent.

Today, we know that time travel need not be confined to myths, science fiction, Hollywood movies, or even speculation by theoretical physicists. Time travel is possible. For example, an object traveling at high speeds ages more slowly than a stationary object.[1] This means that if you were to travel into outer space and return, moving close to light-speed, you could travel thousands of years into the Earth's future. In addition to high-speed travel, researchers have proposed numerous ways in which time machines can be built that do not seem to violate any known laws of physics. These methods allow you to travel to any point in the world's past or future and are discussed toward the end of this book.[2]

Newton's most important contribution to science was his mathematical definition of how motion changes with time (Fig. 1). He showed that the force causing apples to fall is the same as the force that drives planetary motions and produces tides. However, Newton was puzzled by the fact that gravity seemed to operate instantaneously at a distance. He admitted he could only describe it without understanding how it worked. Not until Einstein's general theory of relativity was gravity changed from a "force" to the movement of matter along the shortest path in a curved spacetime. The Sun bends spacetime, and spacetime tells the planets how to move. For Newton, both space and time were absolute. Space was a fixed, infinite, unmoving metric against which absolute motions could be measured. Newton also believed the universe was pervaded by a single absolute time that could be symbolized by an imaginary clock off somewhere in space. Einstein changed all this with his relativity theories, and once wrote, "Newton, forgive me."[3]

Einstein's first major contribution to the study of time occurred when he revolutionized physics with his "special theory of relativity" by showing how time changes with motion. Today, scientists do not see problems of time or motion as "absolute" with a single correct answer. Because time is relative to the speed one is traveling at, there can never be a clock at the center of the Universe to which everyone can set their watches. Your entire life is the blink of an eye to an alien traveling close to the speed of light. Today, Newtonian mechanics have become a special case within Einstein's theory of relativity. Einstein's relativity will eventually become a subset of a new science more comprehensive in its description of the fabric of our universe. (The word "relativity" derives from the fact that the appearance of the world depends on our state of motion; it is "relative.")

We are a moment in astronomic time, a transient guest of the Earth. Our wet, wrinkled brains do not allow us to comprehend many mysteries of time and space. Our brains evolved to make us run from saber-toothed tigers on the African savanna, to hunt deer, and to efficiently scavenge from the kills of large carnivores. Despite our mental limitations, we have come remarkably far. We have managed to pull back the cosmic curtains a crack to let in the light. Questions raised by physicists, from Newton to Kurt Gödel to Einstein to Stephen Hawking, are among the most profound we *can* ask.

Is time real? Does it flow in one direction only? Does it have a beginning or end? What is eternity? None of these questions can be answered to scientists' satisfaction. Yet the mere asking of these questions stretches our minds, and the continual search for answers provides useful insights along the way.

Who This Book Is For

This book will allow you to travel through time and space, and you needn't be an expert in physics. Some information is repeated so that each chapter contains sufficient background information, but I suggest you read the chapters in order as you gradually build your knowledge. To facilitate your journey, I start most chapters with a dialog between quirky explorers who experiment with time from within the (usually) safe confines of a Museum of Music in New York City. This simple science fiction is not only good fun but it also serves a serious purpose, that of expanding your imagination. We might not yet be able to easily travel in time like the characters in the story, but at least time travel is not forbidden by the current laws of physics. As you read the story, think about how humans might respond to future developments in science that could lead to time travel.[4]

When writing this book, I did not set out to write a systematic and comprehensive study of time. Instead, I have chosen a selection of topics relating to time travel that I think will enlighten a wide range of readers. Although Einstein's theory of relativity is nearly a century old, its strange consequences are still not widely known. People still often learn of them with a sense of awe, mystery, and bewilderment. Even armed with Einstein's theories, humans have only a vague understanding of time, and various problems and paradoxes still need to be solved.

By the time you've finished this book, you will be able to

- have a good understanding of spacetime diagrams, light cones, and time machines
- impress your friends with such terms as: "cosmic moment lines," "transcendent infinite speeds," "Lorentz transformations," "causal linkages," "superluminal and ultraluminal motions," "Minkowskian spacetimes," "Gödel universes," "closed timelike curves," and "Tipler cylinders"
- write better science-fiction stories for shows such as *Star Trek* or *The X-Files*
- write computer simulations for various aspects of time travel
- understand humanity's rather limited view of time
- understand that time travel is possible

You might even want to go out and buy a Chopin recording.

Time

A Traveler's Guide

Prelude

Chopin's music is so beautiful that it is hard to think of him as one of the most radical innovators in musical history. But he was none the less a revolutionist for being a gorgeous one; no less an earthquake for not wrecking buildings. He certainly altered the landscape and the very harmonic materials of music. It would be impossible to say which of his works are best where almost all are best: nocturnes, ballads, valises, preludes, polonaises. He gave his dances passionate meaning, and even his finger exercises, or etudes, are more important than the iceberg symphonies of most composers.

—*The Volume Library*, 1928

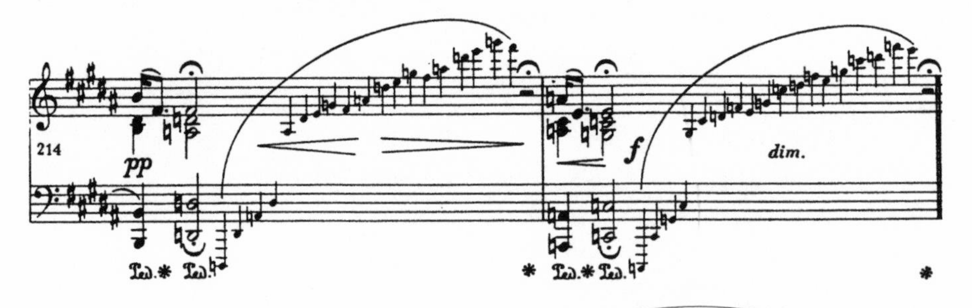

While every one of us is a time traveler, the cosmic pathos that elevates human history to the level of tragedy arises precisely because we seem doomed to travel in only one direction—into the future.

—Lawrence M. Krauss, *The Physics of Star Trek*

Music and higher mathematics share some obvious kinship. The practice of both requires a lengthy apprenticeship, talent, and no small amount of grace. Both seem to spring from some mysterious workings of the mind. Logic and system are essential for both, and yet each can reach a height of creativity beyond the merely mechanical.

—Frederick Pratter, 1996

Prelude

✍ The year is 2063, and you are chief curator of the Museum of Music located on Fifth Avenue in New York City. The museum contains instruments and musical artifacts from various cultures. Even the smell of the place adds to its sense of enchantment—a smell of dust and relics and secrets. Because it is evening, the museum is nearly empty. You are comforted by piano music that begins to quietly pour from the museum's intercom.

You turn to your assistant while pointing to a speaker mounted on the wall. "Mr. Veil, that haunting nocturne is by the great Frederic Chopin."

Your assistant is a Zetamorph, a member of a race of philosophers from a subterranean air pocket on Ganymede, a moon of Jupiter. Mr. Veil has the beautiful warm eyes of Robert Redford, but unfortunately that is where the similarity ends. Above his eyes is a deep, fleshy, head shield, shaped like a helmet. His body consists of numerous segments diminishing in size toward his rear, which is about the size of your little finger. Two whiplike extensions are attached to this last segment and protrude from his ultra-tiny rump.

"The music—it's beautiful," Mr. Veil says. His slight hesitation is betrayed by the popping of tiny air sacs along his forelimbs. Perhaps he feels you are about to send him on a dangerous expedition.

You tap on the speaker. "Chopin's piano-playing was as exquisite as his music. His nineteenth-century contemporaries called his performances 'irresistible,' and his playing was the last word in delicacy and refinement."

Mr. Veil sighs. "Seems to lose something over the speaker system."

"Doesn't matter. We'll get to hear him play."

"What?"

"We're going back in time."

The Zetamorph begins to tap on the floor with his right forelimb. "Sir, time travel—it's impossible."

"It's possible, and I'm going back." You take a step closer to Mr. Veil. "Interested in coming?"

The Zetamorph takes a deep breath. "Sounds safer than your last idea of sending me into the New York City subway."

You raise your left eyebrow. "Certainly. What could go wrong?" ✍

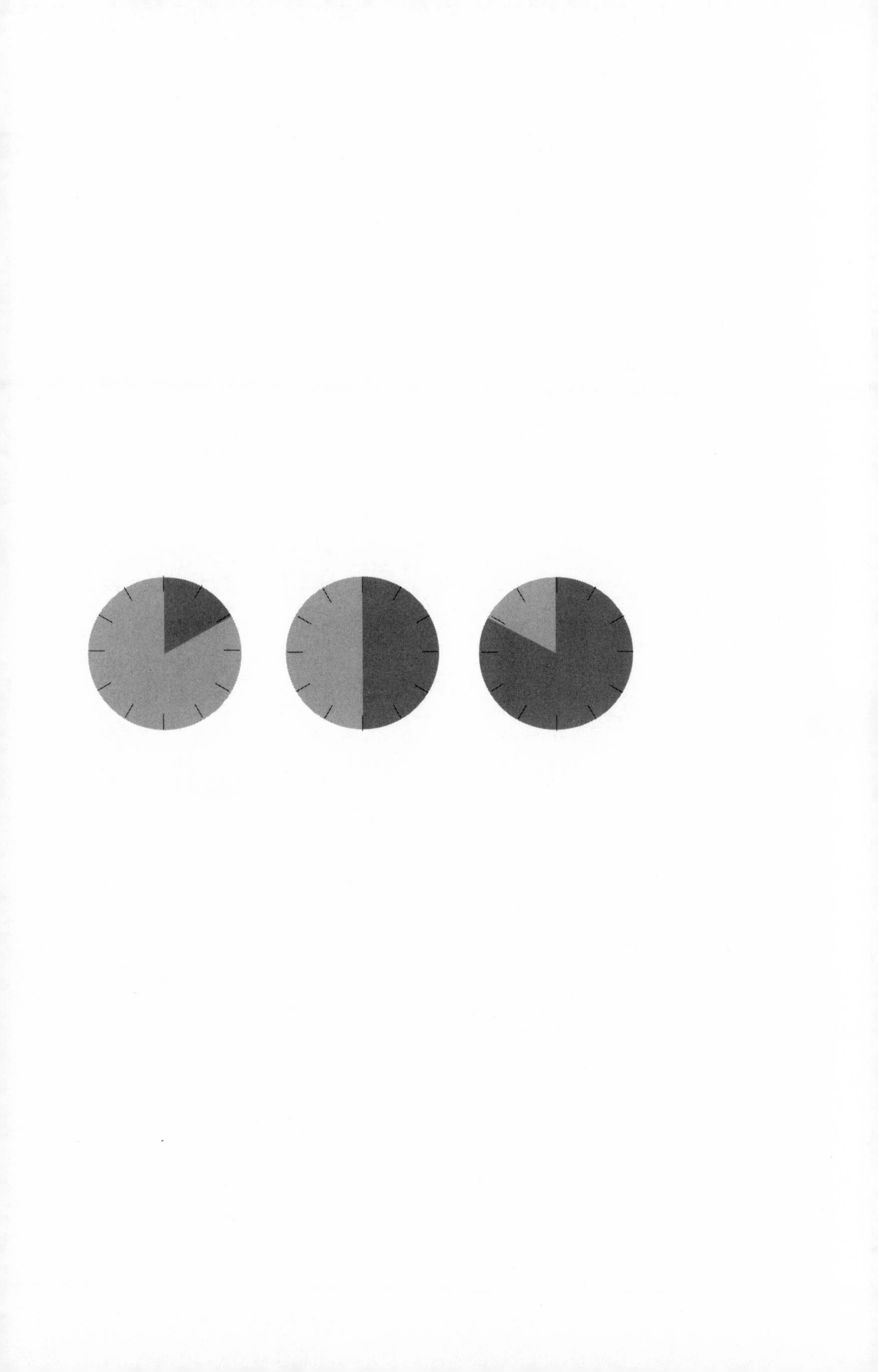

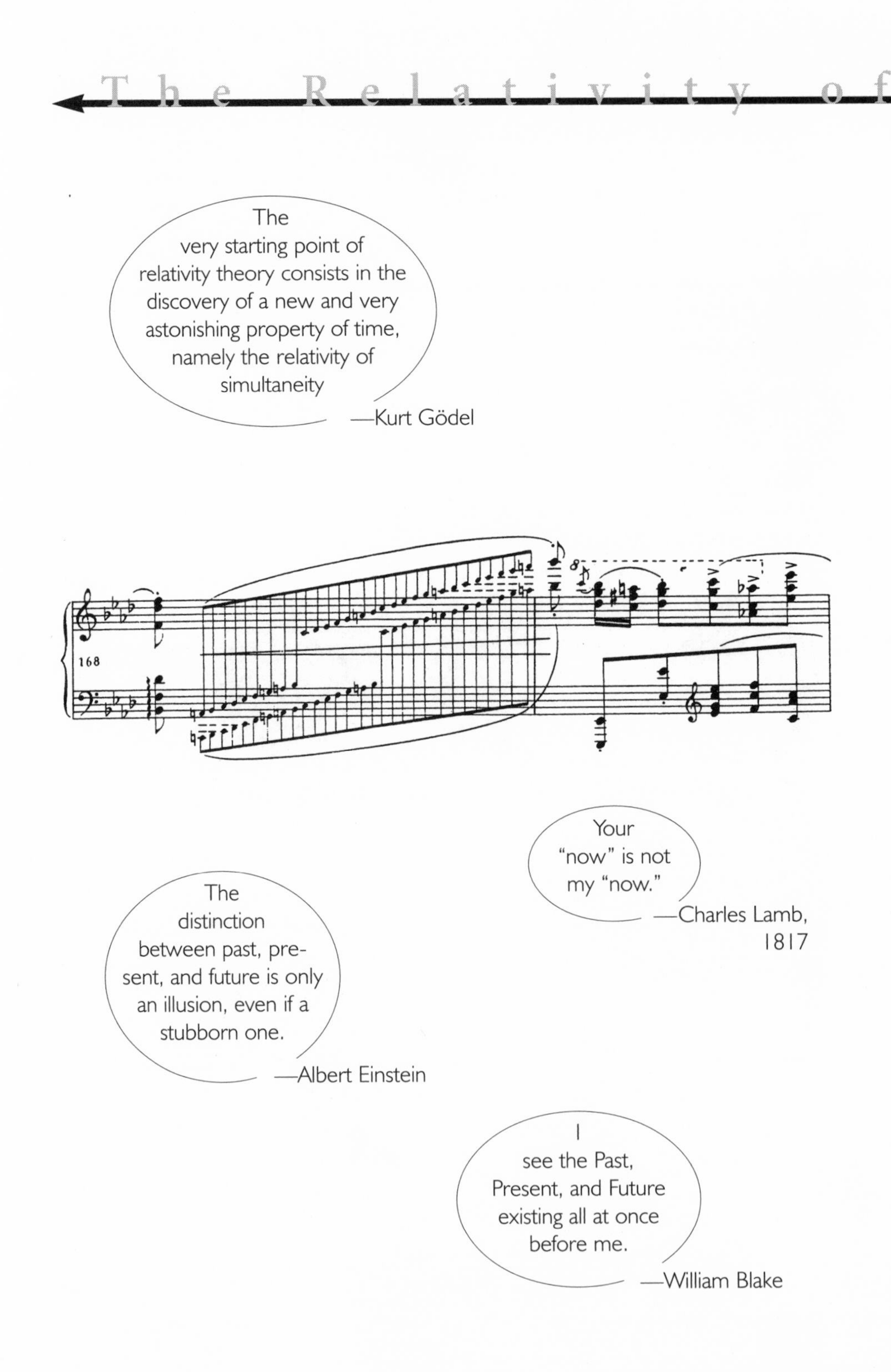
The Relativity of
The very starting point of relativity theory consists in the discovery of a new and very astonishing property of time, namely the relativity of simultaneity
—Kurt Gödel
168
8
Your "now" is not my "now."
—Charles Lamb, 1817
The distinction between past, present, and future is only an illusion, even if a stubborn one.
—Albert Einstein
I see the Past, Present, and Future existing all at once before me.
—William Blake

The Relativity of Simultaneity

1

✍ "Mr. Veil, what the hell are you doing?" You put down a cup of steaming coffee and sit on a bench in the museum's outer courtyard. Several trochophores—slender, winged, devil-faced humanoids—are watching you and Mr. Veil with exponentially increasing interest. They're harmless when they've been detongued and neutralized.

Mr. Veil removes a bassoon from his strangely articulated mouth. "Sir, I was just trying to play—"

You jump up and snap your fingers. "We've got work to do. I want to test some theories about time before we go back and listen to Chopin in the flesh."

To enter the Museum of Music in New York City, you first cross an eighteenth-century courtyard beside an old abbey church. As you enter, you see various instruments lining the long hallway. Some of the instruments, like the cellos, are intact though peeling and corroded by the gentle acid of time. Others are only skeletons—the neck of a viola here, the bridge from a guitar there—as if they all have suffered some indescribable torture.

Mr. Veil quickly comes forward and whips out a thumbnail computer from a nearby nook. "Excellent!" he says.

"You won't be needing that today. I want you to go out and perform a few experiments. Take the car."

Mr. Veil takes a step backward. "Is this safe?"

You wave your hand. "Nothing to worry about. Take two hand-held lasers."

Mr. Veil eyes you suspiciously as he grabs two lasers and leaves for the car. As you wait, you press a button on the wall, causing the museum's computer to play one of Chopin's mazurkas. Thanks to skillful programming, the computer can play all of Chopin's mazurkas, polonaises, and nocturnes, which are stored in the computer's memory. The computer simultaneously projects gorgeous, electro-improvisational holograms all around you. You begin to sway with the music as Mr. Veil leaves the museum.

New York City law says that sports cars must stay grounded inside city limits, but you can see Mr. Veil lift the car's nose before it has moved at all, and the car is pointing straight up as he clears the parking lot. Half a click straight up, while his gee meter climbs—two, three, four—he cuts the power and lets the car glide with its wings opened a trifle.

"Mr. Veil, the car is new, and I've installed some of my own custom features. Be careful."

"Sir, you can relax. I've taken all the driving classes that the police take. The car's in safe claws. All you have to worry about are the New York drivers."

"Very reassuring," you say with heavy irony.

"Ready," Mr. Veil radios to you.

You wave back as his car glides away. "Good, Mr. Veil. Now I want you to find the piles of magazines and newspapers that I left in the car. Place the newspapers in the front of your car. Place the magazines at the rear."

Mr. Veil gasps. "Sir, what shall I do with the *National Enquirers* and *Playboys*?"

"Please do as instructed."

A few minutes later, Mr. Veil says, "Ready. The magazines are in the rear. Newspapers in front."

"Good. Now stand precisely in the middle of the car, and aim one of your lasers at the newspapers, and the other laser at the magazines."

"Ready."

"Mr. Veil, now instruct the car to travel at a constant velocity along-

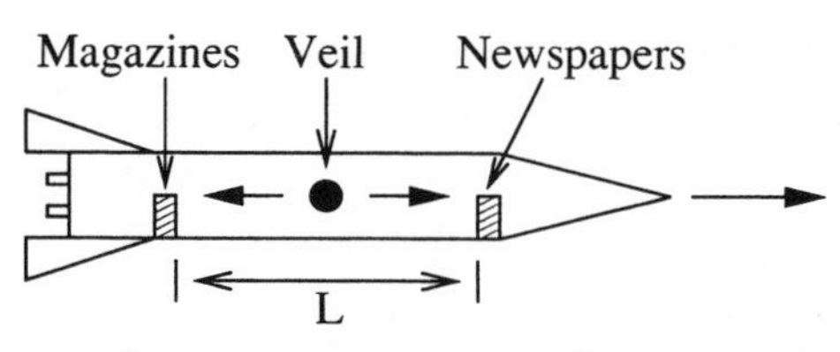

Figure 1.1 Mr Veil firing a laser at newspapers and magazines in order to test the "relativity of simultaneity."

side the museum, and the moment you see me I want you to fire at the pile of magazines and the pile of newspapers."

Soon you see Mr. Veil's car, and just as he passes you he fires at both piles (Fig. 1.1).

"Sir, both piles are on fire, as instructed."

"Good. Now here's the hard question. Which pile ignited first?"

"Sir, what about the fires?"

"Pay no attention to them. The inner lining of the car is coated with flame-retardants."

Mr. Veil coughs as smoke begins to fill his car. "Sir, your question is too easy. I fired on the piles at precisely the same instant. My lasers hit the piles at the same time. The distance L between piles is 20 feet, and I am 10 feet away from both piles."

You raise the volume of Chopin's mazurka. "Please continue, Mr. Veil."

"My two laser beams, each traveling at the speed of light c, took the same time to reach each pile. Because time equals distance divided by speed, each pile ignited $10/c$ seconds after I triggered the lasers. It takes the light from the fire an additional $10/c$ seconds to reach my eyes, so I saw both fires start at the same time, $20/c$ seconds after I fired. They ignited at the same time."

You nonchalantly flick your Calvin Klein infrared glasses and say,

"You're dead wrong."

"What?"

"Your car moved from left to right at a constant speed past my museum."

"Agreed."

"Your forward-traveling laser beam chased after the receding newspapers. Your rearward beam traveled toward the magazines which were approaching the beam. This beam got to the magazines first because the beam traveled less distance then the forward-moving beam, which had to catch up with the newspapers. Therefore, I saw the magazines burst into flame *before* I saw the newspapers catch fire."

"No way!" Mr. Veil says.

"Remain calm, Mr. Veil. This is your first lesson—the relativity of simultaneity. It's the most important lesson for time travellers. For you, the events were simultaneous, but for me the magazines burst into flames first. This means that our relative motions alter the concept of simultaneity. Our two 'nows' are different. Most of us grow up thinking that reality is in the present, and that time can be divided easily into past, present and future. But we should throw out such beliefs. Mr. Veil, do you realize this is the most significant contribution to interplanetary peace since World War III?"

Mr. Veil takes a deep, raspy breath. "Can this help us travel in time?"

"Perhaps. Let's go over the basics of time travel before going back in time to hear Chopin. Tomorrow, we'll discuss time dilation."

"Oooh! Dilation! Mr. Veil's whip-tails tremble. "I can't wait."

You feel pleasantly aroused, having titillated the Zetamorph. "Mr. Veil, get a hold of yourself." You raise the volume of the Chopin mazurka even further, and then say, "Please bring the car back. But first get to work and grab a fire extinguisher." ✍

The Science Beyond the Science Fiction

> How can the past and future be when the past no longer is and the future is not yet? As for the present, if it were always present and never moved on to become the past, it would not be time but eternity.
>
> —Augustine

> The most insignificant present has over the most significant past the advantage of reality.
>
> —Arthur Schopenhauer

Virtually all physicists would agree with the results of the time experiment that you conducted with Mr. Veil. You saw the magazines ignite first for the reason described in the chapter, and also because the light from the magazine fire had less distance to travel to your eye. However, according to Mr. Veil, the piles of newspapers and magazines ignited at the same time. There is no universal "Now." You and I can watch the same physical process and yet disagree on the time when various events took place. Your present is not my present. Your Now is not my Now.

The relativity of simultaneity is one of the most important and profound concepts of physics, and a revolutionary insight into the nature of time. The notion of an absolute cosmic time, with absolute simultaneity between distant events, was swept out of physics by Einstein's equations dealing with the nature of space and time. In fact, it makes little sense to ask questions like, "What happened on a distant star at the precise moment President Kennedy was killed?"

By changing your state of motion, you cause a mismatch of "nows" in terms of what you deduce is happening at the moment. The effect increases with distance. Suppose you get up from your chair and move forward six feet. You have just changed simultaneity with an event in the next galaxy by an entire day. The distant event jumps either into the future or past depending on whether you walked toward or away from the distant galaxy. *The time order of an event on Earth and on another galaxy can be reversed.* However, for this to happen, the two spatially separated events must occur sufficiently close in time so that light (or any signal) doesn't have time to get from one event to the other. As a result, there can be no causal connection between the two events because no information or physical influence can travel faster than light between the events. This means that the time order of two events can be changed at whim simply by ambling about, but you can't reverse cause and effect, producing causal paradoxes. For example, different observers may disagree about whether Mr. Veil's laser beam reached the two ends of the ship at the same time, or if the magazines are hit before the newspapers; but all observers agree that the blasts from the lasers leave the lasers before they arrive at the newspapers and magazines. Paul Davies, author of *About Time,* gives a particularly colorful description on the relativity of simultaneity:

> If reality really is vested in the present, then you have the power to change that reality across the universe, back and forth in time, by simple perambulation. But, then, so does an Andromedan sentient green blob. If the blob oozes to the left and then the right, the present moment on Earth (as judged by the blob, in its frame of reference) will lurch through huge changes back and forth in time.

What the relativity of simultaneity means is that events in the past and future are as "real" as events in the present. As you proceed through this book, you'll find that dividing reality into past, present, and future is meaningless from a physical standpoint. In fact, when mathematical physicist Alfred A. Robb (1873–1936) first heard about the relativity of simultaneity, he wrote, "This seemed to destroy all sense of the reality of the external world and to leave the physical Universe no better than a dream, or rather a nightmare."

Our physical science does not necessarily deal with reality, whatever that is. Rather, it has merely generated a set of consistency relationships to explain our common ground of experience. . . . We have developed these mathematical laws based ultimately on a set of definitions of mass, charge, space, and time. We don't really know what these quantities are, but we have defined them to have certain unchanging properties and have thus constructed our edifice of knowledge on these pillars.

—William A. Tiller, 1976

Science advances but slowly, and with halting steps, But does not therein lie her eternal fascination? And would we not soon tire of her if she were to reveal her ultimate truths too easily?

—Karl von Frisch

It is impossible to meditate on time and the mystery of the creative passage of nature without an overwhelming emotion at the limitations of human intelligence.

—Alfred North Whitehead, *The Concept of Nature*

Building an Einstein-Langevin Clock

2

✍ "Who is that?"

Mr. Veil screams in a metallic voice.

You smile. "I'd like you to meet my new assistant. Constantia Gladkowska from Moscow University. IQ 160. Expert on the psychoacoustics of cello strings. She got her musical education at the Lyceum of Warsaw."

You are in the hall of antique stringed instruments and surrounded by all manner of violins, cellos, and oddly shaped devices from Alpha Centauri. This wing of the museum was originally conceived as a priory, and years later as a revolutionary hall of arcane musical devices. Wealthy benefactors living in Trump Tower continue to give financial support for keeping the wing open to the public.

Over the museum's intercom comes a Chopin polonaise.

Mr. Veil hesitantly ambles to the young woman and extends his forelimb. He seems rail thin, cadaverous beneath an elegant exoskeleton that glistens like rubies.

"Pleased to meet you, Miss Gladkowska," he says smiling with dozens of teeth. Then he whispers in your ear, "Sir, I thought *I* was your assistant."

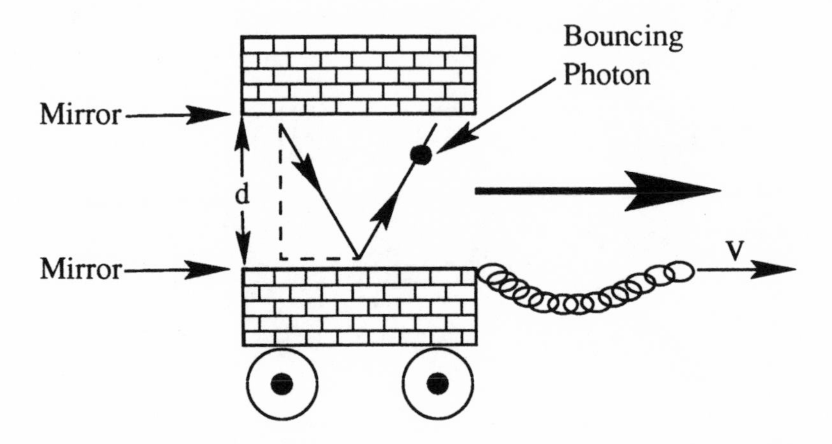

Figure 2.1 A photon clock on wheels.

"Mr. Veil, don't worry, I'll always need you for the dangerous experiments."

Mr. Veil, his whip-tails still dirty with soot from the magazine fire, backs up and nearly overturns an antique cello from the twentieth century.

Constantia draws her fingers through her hair, which cascades over a cloak of tawny orange suede. She is a slim woman, in her mid-twenties, with honey-colored hair and eyes of luminous jade. As she walks forward, her plastic/platinum hybrid boots make a tiny clicking noise on the floor. "Mr. Veil, don't worry; I'm here to learn, just like you. Maybe someday I can go back in time and visit with Peter the Great, the most outstanding of the Russian czars." She gives his forelimb a squeeze and looks at him with a charming smile that involves her eyes as well as her mouth. The Zetamorph walks away with his rump tightly coiled against his body. He stands and squints at you.

You walk over to a chalkboard. "Today's lesson is on *time dilation* and *photon clocks*." As you say the words "time dilation" you think you see a special sparkle in Constantia's eyes, and Mr. Veil makes a groan-like, gutteral-reflex noise.

You reach into a cabinet and bring out a device consisting of two parallel mirrors (Fig. 2.1). "This is called an *Einstein–Langevin clock* or *photon clock*. Now, I want you both to imagine a particle of light, a photon, bouncing back and forth between the mirrors. Can either of you tell me how long it takes a particle to bounce back and forth?"

Mr. Veil raises his forelimb.

You look into his eager eyes. "Yes, Mr. Veil?"

Mr. Veil grabs a piece of chalk. "Sir, an object requires one hour to move 10 miles at a speed of 10 miles per hour—because time equals distance divided by speed. If d is the distance between mirrors, the time t' for the photon to bounce from one mirror to the other and return is $2d/c$."

You motion for Mr. Veil to put down the chalk. "Good Mr. Veil, but I'm the one who uses the chalk around here."

Next you put the clock on a little wooden wagon and drag it across the table. "As the clock travels, what path does the photon take?"

Constantia raises her hands, and you notice that her fingernails are blinking. Latest fad. Probably bioluminescent with luciferin from firefly extracts. You look into her large eyes. "Constantia, you don't have to raise your hand. We're informal here. Please, just speak."

She nods. "Sir, the photon is no longer traveling in a straight up-and-down line. As we watch the moving clock, the photon traces out a zigzag path." She traces an up-and-down line in the air with her finger.

You hit a button on the wall and say, "Let's get in the mood." Chopin's *Fantaisie in F Minor* pours from the speaker.

Constantia smiles. "Ooh, I love that piece. Did I tell you my family loved Chopin's music?"

Your heartbeat increases, but you try to shield your excitement from her. "Right, the path is triangular in our frame of reference." You sketch a zigzag path on the board.

Mr. Veil is deep in concentration. "Sir, since the photon travels a greater distance when the clock is moving, we should observe that a bounce of the photon from top to bottom back to top should take more time than when the clock was stationary."

You sketch further on the board. "Excellent, Mr. Veil! In fact, we can now determine the new path length back and forth by computing the length of the hypotenuse of a right triangle and then doubling it:

$$path = 2\sqrt{d^2 + (vt/2)^2}$$

Here v is the velocity at which I pull the clock along the table and t is the time required for the photon to bounce from top to bottom to top according to our *stationary* frame of reference. We can solve for t, the time the photon takes as measured by us:

$$t = 2\sqrt{d^2 + (vt/2)^2}/c$$

We can combine this equation with the previous one ($t' = 2\ d/c$) describing the tick interval for an observer in the same frame as the clock, and get

$$t = \frac{t'}{\sqrt{1 - (v/c)^2}}$$

The variable t is the time required for a photon to go back and forth as measured by us.

Constantia comes closer. "Sir, I notice that when you stop moving the photon clock, and $v = 0$, the formula reduces to $t = t'$."

"Yes, as expected, the time for the photon trips are equal when there's no motion. Notice that t is always greater than or equal to t', which means we observe the clock to be running slower than a stationary clock. As you move the clock faster and faster, the clock becomes even slower. This is Einstein's *time dilation formula.* What happens when if we move the clock at the speed of light?"

"Great Yggdrasill," screams Mr. Veil. "At the speed of light, where $v = c$, t is infinite. We would observe that it takes an infinite time for the photon to complete it's up and down journey."

"Correct. Time *stands still.*" You perceive a flush on Constantia's face. You imagine she is impressed with your intellectual prowess. But then perhaps it is you who are impressed by her intellectual prowess.

"Sir," she says. "What if the velocity of the clock is greater than the speed of light ($v > c$)?"

You are jolted from your daydream. "If that could happen," you say, "we have a square root of a negative number, and time becomes imaginary. That's one reason why many claim that faster than light travel is not possible. There's also a similar change in length (measured in the direction of motion) that occurs in moving objects. For example, if an object is L' feet long, it shrinks to L feet long according to the formula

$$L = L'\sqrt{1 - (v/c)^2}$$

This means if you move with the object, you will measure its length as L'. However, a stationary observer reports the object contracted to length L. If we were stationary and observed a pencil going close to the speed of light, we would see it shrink to a squashed disk. This is called the *Lorentz–FitzGerald* contraction."

Mr. Veil reaches into a cello and pulls out a notebook computer. He furiously types a computer program (Code 1 in Appendix 2) and says, "Execute." A table of numbers appears on his screen showing the time-slowing factor, $1/\sqrt{1-(v/c)^2}$ for various travel speeds:

(v/c)	Time-slowing factor
0.9	2.29
0.99	7.08
0.999	22.36
0.9999	70.71
0.99999	223.60
0.999999	707.10

"Sir, if something travels 0.999 times the speed of light (99.9% the speed of light), it shrinks 22 times and slows down by a factor of 22."

"Yes, it means the Statue of Liberty (151 feet tall) would appear to shrink to your height (6.8 feet) if we were to launch it at 0.999 times the speed of light. Someone inside the statue's inner chambers would contract by the same amount and also age 22 times slower than us. The slowing we saw for the photon clock doesn't just apply to clocks but to all timing processes."

Constantia's eyes are wide with delight. "The contraction and time dilation effects are small for objects traveling less than half the speed of light, but—" She pauses. "It does mean that at 0.9999 times the speed of light, I would appear squashed to the size of an insect, and saying the single word 'Chopin' would appear to take over a minute."

"Chopin," you say. Your voice is rushed, as you get caught up in the competitive spirit of the conversation. "And if I were to travel at 0.9999 times the speed of light, I could age one year, while you would age 71 years if you stayed where you are and did not go with me." Now your voice is a whisper. "Please, Constantia, we must never let that happen."

Mr. Veil repeats your words very slowly as if imitating the time dilation. *The Fantaisie in F Minor* continues to pour from the museum's speakers, but it seems to be slowing. A fly alights on Mr. Veil's head and then takes off. No one moves. The fly seems suspended, motionless. Reflections from Mr. Veil's body make you feel that you are standing on the periphery of some gigantic crystal as your image is many times reflected.

You feel a chill, an ambiguity, a creeping despair. Constantia and Mr. Veil are still. Their eyes are bright, their smiles—although very unsimilar—are relentless and practiced. For a moment, the cellos in the room seem to glow. But when you shake your head, the glow is gone.

You shake your head again, and the world is back to normal. Perhaps you are just nervous about your plan to create a time machine and go back. There are dangers. Perhaps you have gone too far with Constantia. What is too far when you feel you've made a fool of yourself?

You wave goodbye, as Mr. Veil and Constantia walk off down a corridor to their respective sleeping quarters. It's time to prepare for tomorrow's lecture on various aspects of the Lorentz transformation. You feel a peculiar sense of deja vu. . . .

Yes, it's been a long day. You lie in bed listening to the sounds of crickets coming in through the window, so close they sound as if your room has been magically transported to the countryside. You even smell grass—a fruity aroma that seems out of place in the museum. You are drifting off to sleep, and as you do, you whisper the word "freedom" and dream about photon clocks, time dilation, Lorentz–Fitzgerald contractions, and preparations for your trip back in time. ✍

The Science Behind the Science Fiction

> "The possibilities of space travel beckon us every time we gaze up at the stars, yet we seem to be permanent captives in the present. The question that motivates not only dramatic license but a surprising amount of modern theoretical physics research can be simply put: Are we or are we not prisoners on a cosmic temporal freight train that cannot jump the tracks?"
>
> —Lawrence M. Krauss, *The Physics of Star Trek*

A Very Brief History of Time

The mathematics of time dilation discussed in this chapter could have been studied centuries before Einstein because no special mathematics or physics is required. However, it was not until the twentieth century that the major revolutions in our understanding of time took place.

For thousands of years, before mathematics, protohumans probably had only dim notions of time: past, present, and future. Gradually, humans learned to use various heavenly bodies—for example, the moon, planets, sun, and other stars—to provide a means for measuring the passage of time. Ancient civilizations relied upon the apparent motion of these bodies through the sky to determine seasons, months, and years. We know little about timekeeping in prehistoric eras, but wherever archeologists find records and artifacts, the archeologists also often find evidence of cultures preoccupied with measuring and recording the passage of time. Ice-age hunters in Europe over 20,000 years ago scratched lines and chiseled holes in bones and sticks, possibly counting the days between phases of the moon. Five thousand years ago, Sumerians in the Tigris–Euphrates valley in today's Iraq fashioned calendars that divided the year into thirty-day months, divided the day into twelve periods (each corresponding to two of our hours), and divided these periods into thirty parts (each like four of our minutes). Stonehenge architects who aligned massive columns of stone over 4000 years ago in England left no written records; however, the alignments of stones and holes suggest that Stonehenge could be used to determine seasonal or celestial events such as lunar eclipses and solstices.

Other cultures also had early calendars. Egyptians first based their calendars on the lunar cycles, but around 4235 B.C., they used the star Sirius (the "Dog Star"), which rose next to the sun every 365 days, to create a 365-day calendar. Some scholars set "4235 B.C." as the earliest recorded year in history.

Before 2000 B.C., the Babylonians (again in Iraq) had a 354-day year based on twelve alternating twenty-nine-day and thirty-day lunar months. The Mayans of Central America (2000 B.C.–A.D. 1500) relied on the sun, moon, and Venus to establish 260-day and 365-day calendars. Their written celestial-cycle records indicated their belief that the world was created in 3113 B.C. Mayan calendars later became portions of the great Aztec calendar stones. Other civilizations, such as our own, use a 365-day solar calendar with a leap year occurring every fourth year.

Figure 2.2 The first clocks had no minute hands; in fact, the minute hand gained importance only with the evolution of modern industrial societies. During the Industrial Revolution, trains began to run on schedules, factory work started and stopped at appointed times, and the tempo of life became more precise. The minute hand had finally become important.

Clocks of All Ages

The first clocks had no minute hands; in fact, the minute hand gained importance only with the evolution of modern industrial societies (Fig. 2.2). During the Industrial Revolution, trains began to run on schedules, factory work started and stopped at appointed times, and the tempo of life became more precise. The minute hand had finally become important.

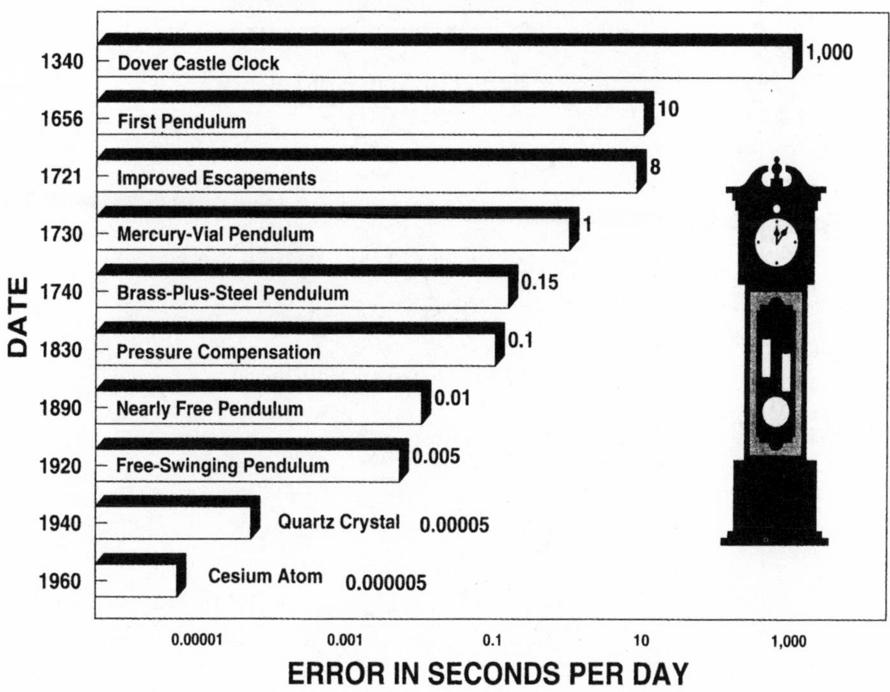

Figure 2.3 The evolution of accuracy in clocks. Early mechanical clocks, such as the Dover Castle, varied by several minutes each day. In the 1980s, cesium atom clocks lost less than a second in 3000 years. Today, an atomic clock known as NIST-7 is accurate to a single second over three million years.

Clocks have become more accurate through the centuries (Fig. 2.3). Early mechanical clocks, such as the fourteenth-century Dover Castle clock, varied by several minutes each day. When pendulum clocks came into general use in the 1600s, clocks became accurate enough to record minutes as well as hours. In the 1900s, vibrating quartz crystals were accurate to a few ten-thousandths of a second per day. In the 1980s, cesium atom clocks lost less than a second in 3000 years, and, today, an atomic clock known as NIST-7 is accurate to a single second over 3 million years. (It is accurate to 0.00000000000001 second per second. For more information on atomic clocks, see note 1.)

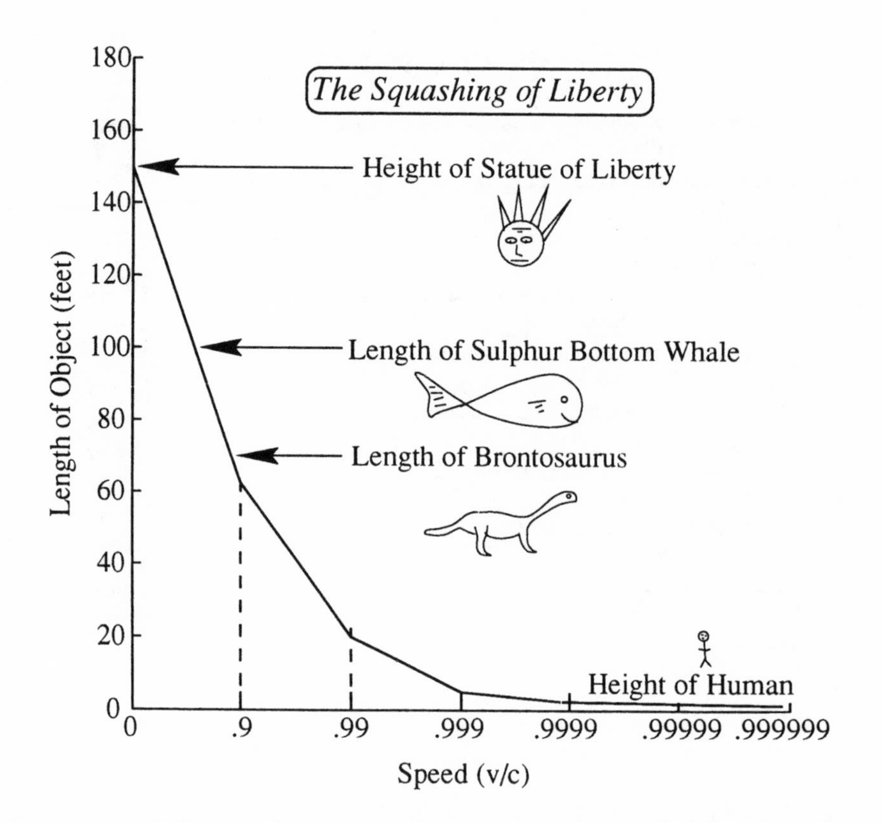

Figure 2.4 The Statue of Liberty contracts in length by traveling headlong close to the speed of light.

Imagine what our world would be like today if, for some reason, our clocks were no more accurate than the Dover Castle clock. What scientific and sociological effect would this have? As just mentioned, minute hands gained importance only during the Industrial Revolution. Imagine a modern world with clocks lacking minute hands!

Crush Liberty Using Your Personal Computer

Code 1 in Appendix 2 should let you experiment with both length[2] and time contraction using the Lorentz formula. For example, I used the program to compute the "Crushing of Liberty" (Fig. 2.4). The Statue of Liberty is 152 feet tall at rest and shrinks as it travels, headfirst, closer and closer to the speed of light. The computer program successively computes the length of the statue as it goes faster and faster. The ratio of the statue's speed to the speed of light is computed by adding "9s" to the left of the decimal; for example $v/c = 0.9, 0.99, 0.999$, and so on. At 0.999 times the speed of light, the statue is shrunk by a factor of about 22, and so the Statue of Liberty would be about as tall as a human. Inside the statue, time is slowed by the same factor relative to an outside stationary observer. How fast would you have to go to crush Liberty to the thickness of a penny?

Imaginary time is another direction of time, one that is at right angles to ordinary, real time. We could get away from this one-dimensional, linelike behavior of time. . . . Ordinary time would be a derived concept we invent for psychological reasons. We invent ordinary time so that we can describe the universe as a succession of events in time, rather than as a static picture, like a surface map of the earth. . . . Time is just like another direction in space.

—Stephen Hawking, *Playboy,* 1990

What I'm really interested in is whether God could have made the world in a different way; that is, whether the necessity of logical simplicity leaves any freedom at all.

—Albert Einstein

The supreme task of physics is to arrive at those universal elementary laws from which the cosmos can be built up by pure deduction. There is no logical path to these laws; only intuition, resting on sympathetic understanding of experience, can reach them. . . . this is what Leibniz described as "pre-established harmony."

—Albert Einstein

The art of conversing with stones is called physics. The question-and-answer periods of the conversations are called experiments. . . . The language spoken is mathematics.

—J. T. Fraser

The Lorentz Transformation

3

✍ "Mr. Veil, I have to go back!"

"Back where?"

"There's no debating it. I have to go back and visit Chopin. Look, I found a message coded in his *Ballade in A-Flat Major*."

Mr. Veil intertwines his whip-tails. "What kind of message?"

"It's a code. The beginning notes in the melody create a code for, 'Constantia, I love you.'"

"That's absurd."

You step back and bang your heel into the wall of your small bedroom in the back of the museum. "Mr. Veil, what did you say?"

"Excuse me, Sir. I meant no disrespect. How could he code a message?"

"Starting at middle C, we can associate the next twenty-six notes with the letters of the alphabet."

"I don't know about this. What could it mean?"

"It means that I went back in time and asked Chopin to place the code in his music. Because it is in the piece in front of our eyes today, it means I must go back in time to fulfill

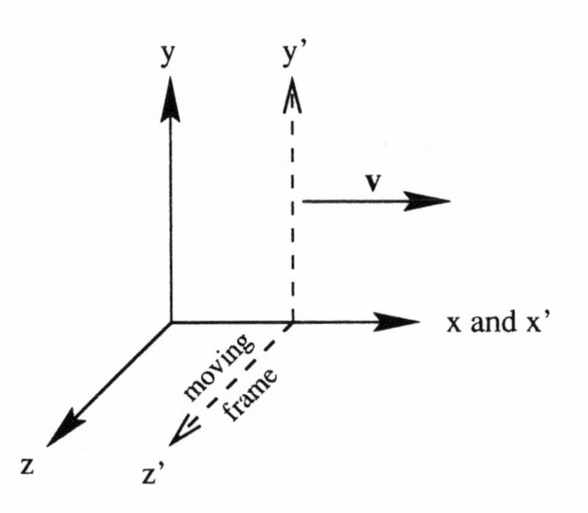

Figure 3.1 Two spatial reference frames in relative motion.

my destiny. I always wanted to go back, but now this means I have no choice."

Constantia enters your room. "What's all the commotion?"

You whisper to Mr. Veil, "Breathe none of this to her." You turn to Constantia. "Ready for the next lesson?"

Around Constantia's neck is a pearl necklace giving her the vague appearance of a young Jacqueline Kennedy. Her shirt is covered with musical notes.

Her shirt's buttons are shaped like little pianos. You stare for a few seconds and wonder where she gets these strange outfits.

"Sir," she says. "I'm ready. These time-travel lessons are wonderful. You should consider writing a book—maybe even a traveler's guide."

You smile. "Today we'll further discuss the Lorentz transformation. Ready?"

You motion Constantia and Mr. Veil to take a seat on your waterbed. There is a vague sloshing sound.

"Let's start by imagining two spatial frames of references.

One frame is stationary. The other is moving at velocity v along the x-axis of the first." You see confusion on their faces. "Hmm, I wish we had a chalk board." In a flash, you reach into Constantia's pocket book, withdraw a lipstick, and quickly make a sketch (Fig. 3.1) on your bedroom mirror. "See?" you say. "It's like sliding one shoebox inside another. We'll use variables with a prime sign to denote the moving frame." You pause. "Now let's place a clock at the origin of each frame. When the origins are on top of one another, we'll set both clocks to time zero, that is, $t = t' = 0$. We can now write four equations that describe how the primed system moves:

$$y' = y$$

$$z' = z$$

$$x' = \frac{x - vt}{\sqrt{1 - (v/c)^2}}$$

$$t' = \frac{t - vx/c^2}{\sqrt{1 - (v/c)^2}}$$

"The stationary observer records (x,y,z) and the moving observer records (x', y', z'). Notice how length contraction takes place only in the direction of motion, and the other spatial dimensions are unaffected. The formulas are called the Lorentz transformations after the Dutchman Hendrik Lorentz, who discovered them in 1904." You pause. "There's another similar-looking equation that shows that the mass of a moving body depends on speed:

$$m = \frac{m_0}{\sqrt{1 - (v/c)^2}}$$

where m_0 is the rest mass when velocity v is 0. We could send Mr. Veil out into the sports car to test these equations."

"Sir," Constantia says, "the mass m of a body becomes infinite at the speed of light when $v = c$. Unless—" She pauses as she puckers her lips.

"Yes?"

"Unless the body is a photon which has zero mass ($m_0 = 0$)—which is probably why you could not accelerate Mr. Veil to the speed of light. To move an infinite mass would require infinite energy."

You smile and raise your right eyebrow. "Constantia, I'm impressed."

Constantia turns away in an expression of ambivalence.

You are worried that she found your remark sexist. "It's just that—"

She turns back. "Been studying time travel in my spare time between cello lessons. Mr. Veil is teaching me to play even better."

You grit your teeth and continue. "Let's next consider two events occurring specifically on the x-axis—for example, Chopin's birth on February 22, 1810, and the hypothetical sneezing of his neighbor a block away. They're simultaneous in the stationary system (at time $t = T$) but are at different places. We'll denote Chopin's location $x = X$ and the sneezer's location $x = X + \Delta X$. "You sketch a diagram on the mirror (Fig. 3.2). "Now let's see what happens if we use the previous equations and consider a moving observer. The occurrence in time of the two events for a moving observer are:

$$t_1' = \frac{T - vX/c^2}{\sqrt{1 - (v/c)^2}}$$

and

$$t_2' = \frac{T - v(X + \Delta X)/c^2}{\sqrt{1 - (v/c)^2}}$$

Now, for the moving observer, we can see that the two events are not simultaneous, being separated in time by

$$t_1' - t_2' = \frac{v\Delta X/c^2}{\sqrt{1 - (v/c)^2}}$$

"Fantastic," Mr. Veil says. "This formula applies to our experiment yesterday. We could have calculated exactly how far apart in time you saw the magazines and newspapers burst into flame." He pauses. "I notice in your equation that only if the two events are in the same place ($\Delta X = 0$) will the two events appear to be simultaneous ($t_1' - t_2'$) to a moving observer."

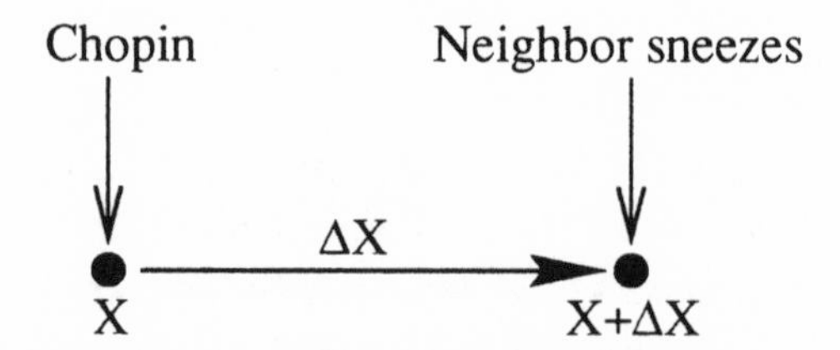

Figure 3.2 Spatial separation of Chopin, located at position $x = X$ and his sneezing neighbor located at $x = X+\Delta X$

You nod. "Notice that the further apart the two events are in space, for example, Chopin's birth and the sneezing neighbor, the further apart in time they would appear to be for the moving observer."

Constantia leans back on your waterbed. "Sir, you've shown us the equations for one frame of reference moving to the right at speed v past a stationary system. But isn't this the same as considering the moving system stationary, and the other system moving to the left at speed $-v$? What do your equations show for this?"

You instruct the museum's computer to play Chopin's *Sonata in B Minor*, and slowly turn to Constantia, as if to heighten the suspense. "It's possible to invert the Lorentz transform, and solve for the unprimed variables in terms of the primed ones, after replacing v with $-v$. It turns out that the Lorentz transform is symmetrical."

Constantia's lipstick is almost finished as you begin to scrawl on the mirror:

$$y = y'$$

$$z = z'$$

$$x = \frac{x' + vt'}{\sqrt{1 - (v/c)^2}}$$

$$t = \frac{t' + vx'/c^2}{\sqrt{1 - (v/c)^2}}$$

Mr. Veil begins to make waves on the waterbed. "Great Yggdrasill! That means that two observers in different frames each say that the other's clock is running slow!"

You nod as you watch Constantia bob up and down on the periodic waves produced by Mr. Veil. "I have a job for you, Mr. Veil. Take the sports car, and radio me when ready."

Mr. Veil jumps from the bed, knocking Constantia to the floor. A few minutes later, he radios you that he is "ready."

"Excellent," you say. "Now I want you to zoom by the museum at high speed. The Earth is our stationary system, and you are the moving system."

There is vague metallic twang to Mr. Veil's voice. "I understand. We'll assume that x and x' are along the direction of motion."

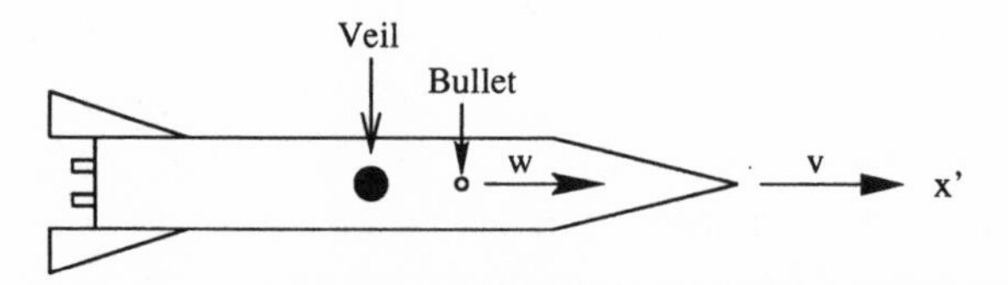

Figure 3.3 Mr. Veil firing a bullet in the space car just as the car races by the museum. The bullet leaves the muzzle at velocity w. The sports car moves with velocity v. How fast is the bullet moving away from the museum?

"Yes. Now I want you to fire a gun toward the front of the car, with the bullet moving at speed w from the muzzle of the gun."

"Sir! I'll destroy the car!"

"Use the Nerf gun in the glove compartment. Its bullets are made of Zandox."

"Excellent, Sir."

"Fire just as you pass the museum."

"Okay, I just fired."

"Mr. Veil. Think hard. How fast did the bullet move away from the museum?"

"Easy. The car is traveling at speed v. The bullet traveled at speed w within the car. Therefore, the bullet moved away from the museum at speed $v + w$" (Fig. 3.3).

"Dead wrong, Mr. Veil."

"Sir?"

"Think of the Lorentz transformation. Inside the car, the position of the Nerf bullet at time t' after you fired the gun is $x' = wt'$. From the inverse Lorentz transformation, the location of the bullet in the museum's frame is

$$x = \frac{x' + vt'}{\sqrt{1 - (v/c)^2}} = \frac{w + v}{\sqrt{1 - (v/c)^2}} t'$$

As you write on the mirror, you see Constantia paying close attention to your every movement, hanging on your every word. She nods enthusiastically as you look at her for confirmation.

You increase the volume of Chopin's *Sonata in B Minor*. "Mr. Veil, now start flying away from the Earth." You pause. "We can use the inverse Lorentz transformation to compute the time t." You write on the mirror with increasing urgency:

$$t = \frac{t' + vx'/c^2}{\sqrt{1 - (v/c)^2}} = \frac{1 + wv/c^2}{\sqrt{1 - (v/c)^2}}\, t'$$

"Therefore the speed of the bullet (x/t) in the museum's frame of reference is

$$\frac{x}{t} = \frac{w + v}{1 + wv/c^2}$$

You hit the mirror so hard with the lipstick that a fragment of it explodes outward.

The car's radio begins to have static. "Sir, may I return now? The oxygen's getting a little low."

"Mr. Veil, please don't interrupt. Notice that for a low-speed bullet (w is much less than c), the result you get from this equation is close to $w + v$, but at high speeds, the situation is very different."

Constantia comes closer to examine the equations, and you notice that she is wearing one of the latest pheromones—probably in a base of hentriacontane to make it particularly potent. "Sir," she says, "what if we imagine that Mr. Veil was firing a laser instead of a gun? The beam of light travels at the speed of light c. In your equation, we'd replace w with c. The old-fashioned assumption is that the velocities should add so that the beam is traveling at $v + c$. This is faster-than-light, or *superluminal,* speed so it must be wrong."

As Constantia says the word "superluminal" you feel your heart begin to race as your inner screams of jubilant exaltation rise to a fever pitch.

Constantia eases the lipstick from your hand and writes with the speed of an alley cat. "I'll use your x/t equation and plug in $w = c$:

$$\frac{c + v}{1 + cv/c^2} = \frac{c^2(c + v)}{c^2 + cv} = \frac{c^2(c + v)}{c(c + v)} = c$$

Therefore, no matter how fast a moving observer goes, he sees the light from the laser traveling at the same speed as does the stationary observer."

"Great Yggdrasill," Mr. Veil screams as he begins to bang on the car window.

"Correct," you say. "This peculiar effect is unique to the speed of light ($w = v$). And look, we didn't have to use any advanced math to figure this out."

"Great Yggdrasill," Mr. Veil chokes.

"Mr. Veil, you may return home now. Slow down. Make sure you have enough atmospheric pressure outside the car, and take a big breath."

You turn to Constantia. "Today we went over several important lessons as a prelude to time travel. You cogitated about the invariance of the speed of light. You pondered how to compute the relative time of events for moving and nonmoving observers, and you learned more about the time implications of the Lorentz equations."

The *Sonata in B Minor* fades from the museum loudspeakers. It is finished. Your heart rate returns to normal. "I suggest we rest now. Tomorrow we have a big day with the brain's internal time machine."

Constantia seems deep in contemplation. After some time she says, "This museum is so wonderful, so spacious. I grew up in a miserable one-room shack in Akademgorodok, Russia. We had no electricity or running water. I hated it. Sometimes tourists came by and took photographs. I felt like I was in a zoo. My mother washed clothes by hand in a twenty-gallon aquarium using water from a nearby gas station."

You turn to her, startled by her sudden and personal outburst. "But I thought your father was a famous professor."

"Yes," she says softly, "but that was much later. He blossomed later in life. When I was growing up, we were dirt poor."

Constantia looks dreamily into your waterbed, and then shrieks, "What is in there?"

"Nothing to worry about. I just adjusted the transparency knob so you could see inside." You smile as you follow her gaze. Inside your waterbed is an oasis. Life is everywhere: beautiful naked gastropods, organisms that resembled land snails, sea snails, limpets, and whelks. It always reminds you of the time you spent exploring the seaports of Galway, Limerick, and Cork on Earth.

You turn to Constantia.

"Care for some sushi?"

"It's late," Constantia says.

Even after Constantia has left the room, the wakes she left in your waterbed continue. You gaze out your skylight. Windows in the rugged skyscrapers form patterns of crystalline checkerboards that scintillate in the dying sunlight. Some of the glass reflects pink beams of light into the heavens. You look toward the sky, always a source of pleasure, with cotton candy clouds, and ebony gaseous mists. Off in the western sky you see chocolate ripples of brown and white amidst twin clouds of flaming incandescent violence. You sigh. New York in the twenty-first century: a nice place to visit, but not a place to spend a lifetime. ✍

The Science Behind the Science Fiction

> Art, as well as science, tries to find meaning; and science fiction in particular tries to make the search rational, understandable, and communicable, by thinking ahead, considering alternatives, and trying to plan for a realistic mental attitude toward change and adaptation."
>
> —Eugene R. Stewart, *Skeptical Inquirer, 1996*

Beings Made of Pure Energy

The change of an object's mass with speed, as discussed in this chapter, was first experimentally observed by the German scientist Walter Kaufmann (1871–1947). In particular, Kaufmann first observed this mass increase for electrons, the lightest particles known aside from massless particles ("luxons") such as the photon and hypothetical graviton, or the neutrino which may have very slight mass. An object's mass increases with increasing velocity, as suggested by the formula:

$$m = \frac{m_0}{\sqrt{1 - (v/c)^2}}$$

where m_0 is the rest mass when velocity v is 0. It turns out that if an electron (which at rest is 2000 times lighter than a proton) is accelerated to near light speed, the electron is measured to carry a momentum equivalent to the protons. If an electron were accelerated to

0.9999999999999999999999999999999999999999999999999999999999999999

times the speed of light, this tiny particle would slam into you with the impact of a Mack truck traveling at 60 miles per hour. This is why it would probably be impossible to accelerate a spaceship, such as the *Enterprise* on

Star Trek, close to light speed and beyond. As Lawrence Krauss in the *Physics of Star Trek* notes, "All the energy in the universe would not be sufficient to allow us to push even a speck of dust, much less a starship, past this ultimate speed limit of light." Because of this mass-variation effect, even when the *Enterprise* is using its "impulse drive" powered by nuclear fusion, each time the *Enterprise* accelerates to half the speed of light, it would have to burn 81 times its entire mass in hydrogen fuel.

I should also point out that not just light but all massless radiation must travel at the speed of light. This means that many beings made of "pure energy"—commonly found in science fiction—must travel at the speed of light. They'd have trouble slowing down, and their clocks would be infinitely slowed compared with our own!

In this chapter, the result derived for the speed of the bullet (x/t) in the museum's frame of reference:

$$\frac{x}{t} = \frac{w + v}{1 + wv/c^2}$$

was first found by the French theoretician Henri Poincaré (1854–1912) before the publication of Einstein's article on special relativity, which also derives these equations showing how velocities add at high speeds. In fact, it was Poincaré who gave the Lorentz transformation its name and who declared that "no velocity could surpass that of light." One of the most important implications of this equation is the invariance of the speed of light in empty space, independent of the observer's frame.

Paradox

I want to digress in the remaining sections of this chapter to give you a background to the physics of time travel so that your strange travels in the rest of the book are more easily understood. Let's get your mind "in gear" by first imagining one kind of time machine. The notion of a time machine is simple enough. Here's one scenario. Imagine that terrorists from outside the United States of America have just bombed the U.S. Capitol Building. You are President of the United States and sitting in your chair in the White House. You place a quick phone call to rogue-warrior Richard Marcinko, and command him to lead his Navy SEALs in a counterattack. You end your phone call at 5:00 p.m. and step into a time machine in the corner of the room that lets you travel back in time by half an hour. A few seconds later (according to your wristwatch) you emerge and find yourself about to place the phone call. The clock on the wall reads 4:30 p.m. Watching from the back corner of the room, you can now wait and watch yourself place the phone call at 5:00 p.m.

The puzzling nature of this situation is familiar to even those who never read science fiction. Consider the President, who returned to the past and is watching the first version of the President. Suppose the returning President trips over the phone cord at 4:45 p.m., or breaks the time machine, or even murders his copy. Then the President could prevent himself from making the phone call: a clear contradiction or paradox. For this reason, some believe that time travel to the past is impossible. But other researchers have provided various mechanisms to avoid paradoxes including the generation of a parallel universe each time a time traveler reenters the time stream.

There are other ways to avoid paradox. For example, the initial President may see a copy of himself standing in the corner of the room, watching, but continue to make the phone call. In fact, the two Presidents might not see each other at all, the older version deciding to live out the rest of his days on some remote Island in the South Pacific. As long as there is at least one nonparadoxical solution, some physicists believe problems involving free will and causality are not generated.

Here's another example of a nonparadoxical solution. One day the President meets a man who looks exactly like himself but who is older. The man tells the President that he is his older self who has traveled back in time. The President thinks the man is crazy and continues with his life. Years later he is compelled to travel back in time to meet his younger self—exactly what his older duplicate had told him when he was younger. Of course, the younger President thinks the older version is insane. The two copies separate until the day comes when his younger self makes the trip back in time. No paradoxes!

What happens if the future President does not instantaneously travel back in time but rather gradually traverses time to visit his former self? This situation gets strange, because it deals with the concept of backward-running time—something first dealt with by philosophers and science-fiction writers.

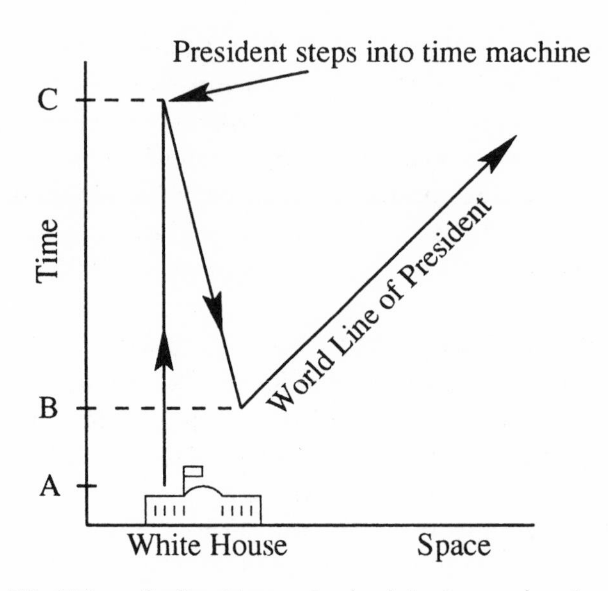

Figure 3.4 World line of a President going back in time and seeing himself.

To help understand it, let's draw a "spacetime diagram" with space on the horizontal axis and time on the vertical. Figure 3.4 shows a zigzagging *world line* of the President. A world line describes the path of an object through space and time. At time *C*, the President goes back to time *B*, possibly converses with his younger self, then continues to lead a normal life. To understand this figure, simply place a straight edge at the bottom of the graph, its edge parallel to the space axis, and move it slowly upward. At time *A* you see the President. At time *B*, an older President suddenly materializes in the White House out of thin air. The older President appears to be living backward. His hair grows shorter with time. Finally, at time *C*, "young" President and "backward" President vanish. The older President continues his life somewhere else in space (at the upper right of the diagram). For an extra copy of the inventor to exist at the same time as the original, it seems reasonable to assume that the time machine will need an input of a great deal of energy, at least equivalent to the mass of the inventor.

Notice that at time *B*, there would actually be *two* copies of yourself moving away in different directions. One of the new Presidents would be moving forward in time, the other backward in time. All together there are three Presidents coexisting. At time *C*, the original President and the reverse President vanish into thin air to leave only a single President (the one moving to the upper right) for all time after *C*. In 1962, Hilary Putnam argued that, although strange, the fact that we can draw these kinds of kinked world lines supports the case for backward time travel—that time travel can make some sense if one considers the possibility of backward-running time.

You've just encountered several time-travel scenarios. We can summarize two main classes of paradox associated with time travel:

- alteration of previously recorded events, for example, going back in time and killing your mother before you were born (this class of paradoxes seems to do the most "damage" to the concept and structure of spacetime).
- the possibility of events without any beginning. For example, consider a struggling author unable to sell his novels to publishers. One day the author meets a wealthy, old gentleman who gives him the secret formulas and futuristic circuitry to make a time travel device. The author goes into the future, steals a few novels from an aging Stephen King and Anne Rice, comes back, and sells the novels under his own name. He becomes wealthy, able to sell anything to Simon and Schuster or Knopf at his slightest whim, and he is loved by his adoring fans. He fires his book agent, having no need for him. Finally, as an old man, the author goes back in time to give his younger self the secrets of time travel. Question: From where did the idea of the time travel device come?

We'll discuss other paradoxes throughout this book.

Relativity Warm-up and Review

Albert Einstein's special theory of relativity is one of humankind's greatest intellectual triumphs.

The theory says that time between two events is usually different when recorded by an Earth observer or a spaceship commander.

Time between events is relative. The theory was derived from two postulates proposed by Albert Einstein when he was only twenty-six:

- Absolute, uniform motion cannot be detected.

- The speed of light is independent of the motion of the source.

From these two simple postulates, which I'll explain shortly, we find surprising results that seem to contradict our common sense. On the other hand, relativity appears to accurately describe nature, from tiny subatomic particles to galaxies. Relativity allows physicists to make accurate predictions about the universe.[1]

The fundamental postulate of Einstein's special theory of relativity is that the speed of light in a vacuum is the same for all observers. This is in contrast to the speed of sound, which changes for an observer moving with respect to the background air. This property of light leads to a loss of simultaneity: two events occurring at the same time as measured by one observer at rest occur at different times for a moving observer—something we've discussed in Chapter 1. Generally, Einstein's theory of relativity emphasizes the notion that no matter what we observe, we always do so relative to a frame of reference that may differ from someone else's, and that we must compare our reference frames to understand our measurements and results about the events we observe.

To best understand the special theory of relativity (which Einstein simply called the Principle of Relativity), consider yourself in an airplane traveling at constant speed. This is called the "moving frame of reference." The Principle of Relativity essentially says that without looking out the window, you cannot tell how fast you are moving. Because you can't see the scenery moving by, for all you know you could be on the ground in a stationary model of an airplane—that is, you could be sitting in a "stationary frame of reference."

All the famous laws of physics are the same in an airplane moving uniformly or in a plane at rest. Also, all the famous numerical constants (such as the elementary charge on an electron or the speed of light $c = 299{,}792{,}458$ meters per second) are the same in all frames in uniform relative motion. However, there are various relationships that can be different for moving and stationary bodies. For example, the time between two events may be different in two different frames of reference. The spatial separation between two events may differ in two different frames.

Einstein's special theory of relativity is also successful in accurately describing what happens when objects move at speeds close to the speed of light. However, the special theory does not explain how gravitation acts on objects. In 1915, ten years after Einstein proclaimed his theory of special relativity, Einstein gave us his *general* theory of relativity, which explained gravity from a new perspective. In particular, Einstein made the startling suggestion that gravity is not really a force like other forces, but results from the curvature of spacetime caused by masses in spacetime.

To best understand this, consider that wherever there's a mass in space, it warps space. Think of a bowling ball sinking into a rubber sheet. It's a convenient way to visualize what stars do to the fabric of the universe. If you were to place a marble into the depression formed by the stretched rubber sheet, and give the marble a side-ways push, it would orbit the bowling ball for a while, like a planet orbiting the sun.

If you rolled a marble on the sheet far from the bowling ball—where there is no funnel in the rubber sheet—the marble wouldn't be affected. If you were too close, it would get sucked into the dent in space. The bowling ball warps the rubber sheet just like a star warps space. Far from the gravitating body, the curvature of the space "funnel" is less pronounced. Physicists describe this with the motto, "Matter makes space bend. Space tells matter how to move."

Even light rays get bent by the curvature of space. Because of this, the apparent positions of stars that we see in the night sky are not always the "actual" positions of the stars.

Time is also distorted in regions of large masses, and in Einstein's theory, the very existence of time depends on the presence of space. Einstein's general theory of relativity can be used to understand how gravity warps and slows time; for example, it explains why time moves slightly slower for you in the basement of your house than on the top floor where gravity is slightly weaker. Not only does general relativity permit time travel, it actually encourages time travel in strange ways that we will discuss later in this book. All of these theories are not speculations, but rather have all been verified by experiments in the lab and in outer space.

Sports Car Analogy

The ideas of special relativity can best be understood with more simple examples. Special relativity begins with the astonishing experimental fact that *c*, the speed of light in a vacuum, never changes, no matter what moving system you measure it from. Consider a useful analogy. Your sports car is traveling at 60 miles an hour. An older car is following you at 40 miles per hour. How fast do you appear to be traveling to an observer in the older car? The apparent (relative) speed of the sports car would seem to be only 20 miles per hour, if you were observing from the older car. But if you could measure the speed of light from the headlights, the rays of light would appear to move at exactly the same speed *c* whether measured from the sports car or the older car. (Why would God make the universe this way?)

The speed of light appears to constrain motions of ordinary objects. What would happen if you lived in a universe where the speed of light were only 60 miles an hour? You'd be quite frustrated, because as your foot pressed the car's gas pedal, you would never quite reach the speed limit no matter how hard you pressed the pedal. Now imagine you are on a high-speed rocket. Real-life rocket ships face the same problem as they approach *c*, the speed of light in a vacuum. As you near light speed, the ship's mass grows toward infinity—meaning that to accelerate all the way to the speed of light, an infinite amount of energy (fuel) is needed. If this frustrates you, at least Einstein's special theory of relativity also says that by traveling closer to velocity *c* you can make your journey seem as short as you like—though only to yourself! Because clocks on the ship run more slowly than stationary clocks on Earth, your long journey appears to you to take a few hours. To those on Earth, your journey may appear to take centuries.

Time is not the only thing that shrinks in your high-speed rocket. As you accelerate, you would see objects outside the ship shrink in length. To you, your flight path and the whole universe seems to shrink along the direction of travel. To you, acceleration does not so much increase your speed as reduce the distance covered. Still the trip is not easy. The energy to move the increased mass of your high-speed ship is almost unimaginable.

Hyperbeings, Hypertime, Eternitygrams, and Musical Scores

> "I take my accurate form," Memnoch continued, "when I am in Heaven or outside of Time."
>
> —Anne Rice, *Memnoch the Devil*

What can hyperbeings do, and why do their acts initially seem so alien to us?

—Theoni Pappas, *More Joy of Mathematics*

One central question of this book is, "Can one build a time machine?" This kind of question can be posed more accurately in the language of Einstein's general theory of relativity, which describes space and time as a unified four-dimensional continuum called spacetime. To best understand this, consider yourself as having three spatial dimensions—height, width, and breadth. You also have the dimension of duration—how long you last. Modern physics views time as an extra dimension; thus, we live in a universe having three spatial dimensions and one additional dimension of time. Stop and consider some mystical implications of spacetime. Can something exist outside of spacetime? What would it be like to exist outside of spacetime? For example, Thomas Aquinas believed God to be outside of spacetime and thus capable of seeing all of the universe's objects, past and future, in one blinding instant. An observer existing outside of time, in a region called "hypertime," can see the past and future all at once.[2] In a strange sense, when we scan back and forth over a musical score we are like a hyperbeing who lives outside of time. In fact, my favorite spacetime analogies are musical scores such as the ones decorating the pages of this book. A musical score makes time solid. A musician can see past, present, and future all at once.

There are many other examples of beings in literature and myth who live outside of spacetime. Many people living in the Middle Ages believed that angels were nonmaterial intelligences living by a time different from humans, and that God was entirely outside of time. Lord Byron aptly describes these ideas in the first act of his play *Cain, a Mystery*, where the fallen angel Lucifer says:

> With us acts are exempt from time, and we
> Can crowd eternity into an hour,
> Or stretch an hour into eternity.
> We breathe not by a mortal measurement—
> But that's a mystery.

As I just mentioned, I like musical scores as a visualization of time. But perhaps a more direct analogy involves an illustration of an "eternitygram" representing two disks rolling toward one another, colliding, and rebounding. Figure 3.5 shows two spatial dimensions along with the additional

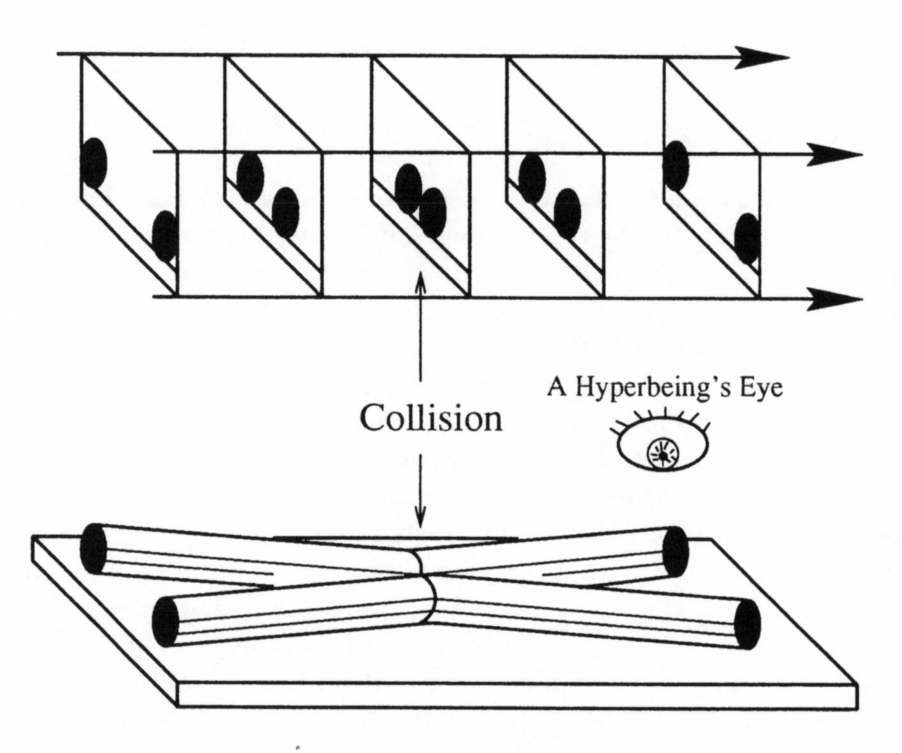

Figure 3.5 An eternitygram for two colliding disks.

dimension of time. You can think of successive instants in time as stacks of movie frames which form a three-dimensional picture in hypertime in the eternitygram. Figure 3.5 is a "timeless" picture of colliding disks in eternity, an eternity in which all instants of time lie frozen like musical notes on a musical score. Eternitygrams are timeless. Hyperbeings looking at the disks in this chunk of spacetime would see past, present, and future all at once. What kind of relationship with humans could a creature (or God) have who lives completely outside of time? How could they relate to us in our changing world? One of my favorite modern examples of God living outside of time is described in Anne Rice's novel *Memnoch the Devil.* At one point, Lestat, Anne Rice's protagonist, says, "I saw as God sees, and I saw as if Forever and in All Directions." Lestat looks over a balustrade in Heaven to see the entire history of the world:

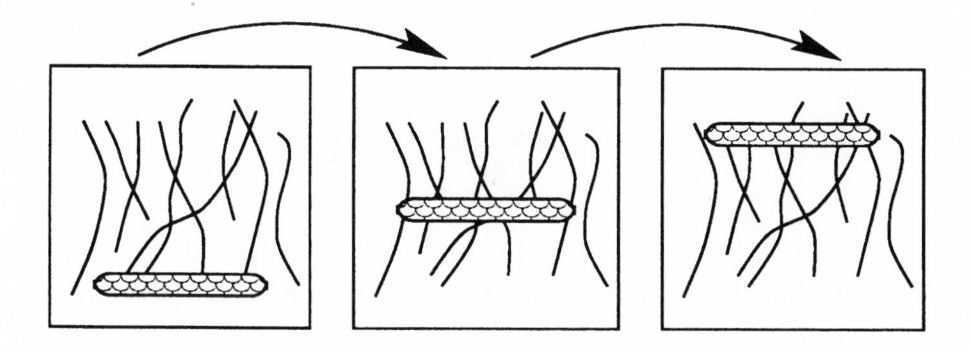

Figure 3.6 Tracks of dancing bees, frozen in spacetime. Is time an illusion? If bee world lines were somehow fixed in spacetime, and all that "moved" was our perception shifting "up the page" as time "passes," we would still see a complex dance of buzzing bees, even though nothing was moving.

> the world as I had never seen it in all its ages, with all its secrets of the past revealed. I had only to rush to the railing and I could peer down into the time of Eden or Ancient Mesopotamia, or a moment when Roman legions had marched through the woods of my earthly home. I would see the great eruption of Vesuvius spill its horrid deadly ash down upon the ancient living city of Pompeii. Everything there to be known and finally comprehended, all questions settled, the smell of another time, the taste of it. . . .

Figure 3.6 shows a schematic diagram of dancing bees in spacetime, just like Figure 3.5 shows two disks colliding in spacetime. The track of each bee through spacetime is represented by a line. In each of the three squares in Figure 3.6, the time axis is vertical. This means that time gradually proceeds up the page, the past at bottom and the future toward the top. The space axis in each square is horizontal. A bee sitting still would be represented by a vertical line in the squares because its horizontal (spatial) component never changes. If all bee world lines were somehow fixed like tunnels in the ice of spacetime, and all that "moved" was our perception shifting up the page in the figure as time "passes," we would still see a complex dance of interacting bees even though nothing was moving. Perhaps an alien would see this differently than us. In some sense, all bee tracks and interactions may be con-

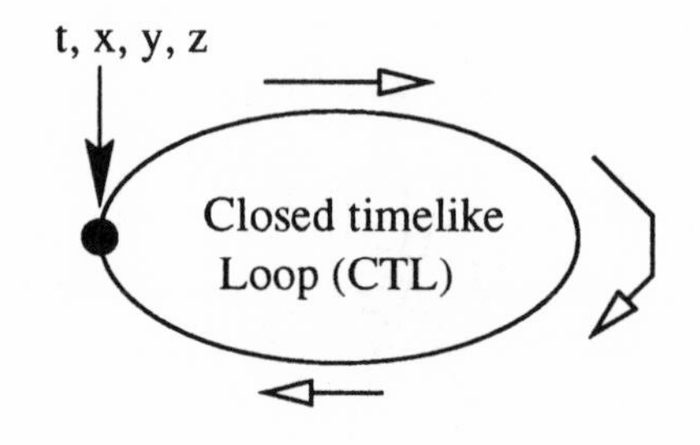

Figure 3.7

sidered fixed in the geometry of spacetime, with all movement and change being an illusion resulting from our changing psychological perception of the moment "now." Some mystics have suggested that spacetime is like a novel being "read" by the soul, the "soul" being a kind of eye or observer that stands outside of spacetime, slowly gazing up along the time axis.

Of course, many stories have been devoted to the experience of time. One of my favorite short stories is Norman Spinrad's "The Weed of Time," which describes a boy who eats a weed that makes him see his whole lifetime as simultaneously present. Therefore, as a baby, he already knows he will eat the weed before (according to our limited point of view) he has ever eaten it! David Masson's "Traveller's Rest" describes a war in a land where everyone's perception of time slows down as they travel south. One young soldier on a short vacation travels south, marries, and has children. Then, in middle age, the soldier receives a message telling him that his vacation is over. He travels north to arrive in his barracks 22 minutes (local time) after he left.

Closed Timelike Loops

Let's close this chapter with a final piece of background information on the physics of time. Consider the term *closed timelike loop*, or CTL, or *closed timelike curve* (CTC), which we'll frequently encounter in this book. To best understand this, consider that all objects in our universe follow a path through spacetime. For example, if you stand very still at a location denoted by $x = 0$, $y = 0$, and $z = 0$, your spacetime path is a line consisting of $x = y = z = 0$ for all values of t.

What spacetime path would a time traveler follow as he went forward in time and then returned to the time from which he started? His path would start at some initial spacetime point defined by one time coordinate and three spatial coordinates (t, x, y, z). The point representing the time traveler would move forward in time (as measured by the traveler's watch) and return to the same spacetime point (t, x, y, z) along some path, or CTL (Fig. 3.7). In other words, a time journey in which you return to the same place you started from at the moment you left is called a CTL. One of the central questions of time travel asks whether it is possible to generate closed timelike curves using ordinary matter moving in ordinary space. We will discuss this precise question in this book.

I'm not sure if others fail to perceive me or if, one fraction of a second after my face interferes with their horizon, a millionth of a second after they have cast their gaze on me, they already begin to wash me from their memory: forgotten before arriving at the scant, sad archangel of remembrance.

—Ariel Dorfman, *Mascara*, 1988

Causes and effects don't seem to fit. Causes and effects are a result of thought. I would think mental illness comes before thought.

—Robert Pirsig, *Zen and the Art of Motorcycle Maintenance*, 1979

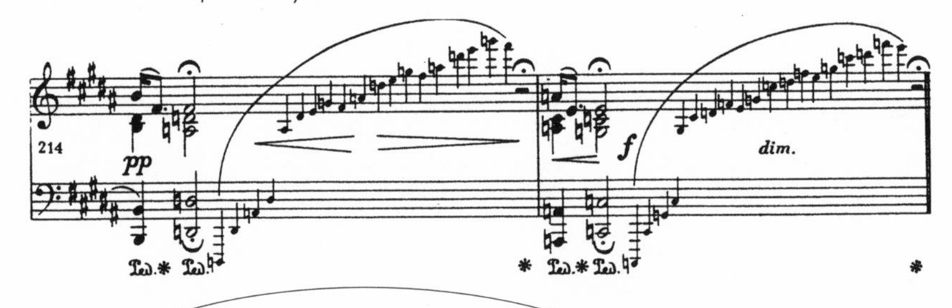

Suppose for example I see one event E precede another, E*. I must first see E and then see E*, my seeing of E being somehow recollected in my seeing of E*. That is, my seeing of E affects my seeing of E*: This is what makes me—rightly or wrongly—see E precede E* rather than the other way round. But seeing E precede E* means seeing E first. So the causal order of my perceptions of these events, by fixing the temporal order I perceive them to have, fixes the temporal order of the perceptions themselves . . . the striking fact . . . should be noticed, namely that perceptions of temporal order need temporally ordered perceptions.

—Hugh Mellor, 1981

Mathematics is the one subject in which time is irrelevant.

—Philip Davis and Reuben Hersh, 1986, *Descartes' Dream*

The Brain's Time Machine

4

✍ "Sir, Constantia is a robot."

"What did you say?"

Mr. Veil sighs. "Not only that, but she's a time traveler from the past, and was once in love with Chopin."

You are in the Hall of Keyboards. All around you are stringed keyboard instruments: clavichords, harpsichords, pianos—even a sixteenth-century clavicytheria. As you jump up, you bang your knee on a Steinway grand piano.

"Mr. Veil, have you lost your mind?"

Mr. Veil's body is glimmering like mercury. "Sir, bear with me. I've done a little research on Chopin. Ancient biographies indicate that he was in love with a Miss 'Constantia Gladowski' in Warsaw in 1830."

A few Zonites with smooth, pulsing bags of flesh hanging along the sides of their faces trudge past dragging duffel bags by their strings. Their bags hiss like snakes as they slide along the parquet floor. Oboes and other wind instruments stick out of their bags.

The Zonites smile blandly at you as they pass by.

You shake your head. "Mr. Veil, it must be a coincidence." You pause. "And why do you think she's a robot?"

"I saw her insert a screwdriver into her right ear to make certain adjustments."

"Mr. Veil, I've known robots, and Constantia is no robot. She must have been using one of those neural stimulators. All the free spirits use them."

"But—"

You snap your fingers. "Mr. Veil, this is crazy. Robots? The 1800s? I think you need a break. Today we won't talk physics."

Constantia walks into the room. On her shirt are several large, blue-and-yellow triangles. Her shirt's button's are shaped like spirals, and her earrings, light-emitting diodes, are blinking. "Hey, you guys whispering about me *again*?" The triangles on her shirt seem to vibrate.

"Mr. Veil has some crazy idea that you're from the 1800s, that you knew Chopin."

"What?"

Mr. Veil towers above Constantia. "Madame, your name—it's the same as one of Chopin's loves."

"Oh, all this whispering is about *that*?" She smiles. "I come from a musical family. They loved Chopin. They named me after Chopin's girlfriend. They liked the name. Nothing strange about that."

You turn to Mr. Veil. "You see. Stop all this nonsense."

Mr. Veil taps his whip-tails on the hardwood floor and says nothing.

You motion for Mr. Veil and Constantia to sit on the piano bench. "Today I want to take a break, switch subjects, and talk about the brain, time, consciousness, and causality. We'll return to pure physics tomorrow. Let's start with several quick experiments with the notion of causality—that cause is followed by effect. Both of you put up your hands whenever you like." You pause and wait, and Mr. Veil and Constantia put up their hands.

"I bet you both think that the movement of your hand is initiated by your willpower. You think that your psychological decision process is the trigger and prime cause of that movement. However, 0.8 seconds *before* you consciously decided to move your hands, there was an electrical signal in your brain called a readiness potential. This has been proven by neuropsychologists. In other words, at the exact instant you *consciously* decided to move your hands, the actions had been already determined almost a second before. By monitoring your brain activity, I could have foreseen your movements before you made your conscious decisions."

Constantia presses a key on the piano keyboard. "Incredible," she says. "You mean a biologist could know that I was about to press this key before I made the conscious decision?"

You nod. "Lesson 1: The relation between cause and effect, as you experience it, does not reflect the actual sequence of causal interdependence." You pause. "Here's another experiment that should disrupt your notions of causality on a human scale. Mr. Veil, come here."

You wave a gleaming electrode in front of him. "I'm going to insert this into your brain."

"What?" he says.

"The electrode is so thin, this won't hurt a bit." Without further hesitation, you insert the wire into his brain. "Mr. Veil, what are you thinking? Let your mind drift."

He looks toward the ceiling as his thoughts drift. "My life on Ganymede. My fellow Zetamorphs. A solitary Ganymedean seabird, wings outstretched and motionless. The bird floats in the drizzle, inches above the limitless icy sea. A pink ray of sunlight reflects off its wings and startles the multitentacular glow-worms drifting on the mist-covered ocean."

"Okay," you say. "This demonstrates another notion of causality and free will. I've been stimulating your brain by the electrodes, and you experience these stimulations as spontaneously arising sensations rather than events I've imposed. Although I'm controlling you, you feel entirely free. Lesson 2: There are important illusions that reflect your brain's *interpretation* of causality."

You bring out another electrode from your pocket. "Now I want to show you that the brain has its own time machine. Constantia, come here."

She looks at the electrode in your hand. "You're not going to stick me with that thing," she says.

You throw the electrode to the floor. "Of course not."

She comes closer. "Okay," she says.

You touch her on her leg. "Don't worry, Constantia. This isn't a clumsy attempt on my part to initiate a relationship with you. When I touch you, your brain needs half a second for computation of the sensory signals. However, you do not perceive the touch half a second later. You immediately experience having been touched. This means that your brain antedated the experience by half a second. Think of your brain as a post office that assigns a date to a letter earlier than the actual date of the letter."

Constantia looks at her leg. "Sir, you mean that my brain knows when the sensory signal arrived, and it compensates for its computation time? It creates the illusion of simultaneity?"

"Correct. Lesson 3: In order to perceive stimuli coming from the outside world in the correct order, we shift the perception backward in perceived time to the moment the stimulation actually had taken place. I call this the brain's time machine."

Constantia looks at you. "Sir, you may remove your hand from my leg."

Mr. Veil taps on the floor with his whip-tails. "Sir, can I remove the electrode in my brain now?"

"Mr. Veil, not yet. For our next experiment I want to measure your reaction time." You hand him a box with a button on it. "Press this button when you perceive my touch." You touch him, and he presses a button.

"Sorry, Sir. I pressed the button by accident."

"You did not. Here's what happened. Due to normal reaction time, you pushed the button 0.2 seconds after I touched you. However, 0.3 seconds later, I retroactively masked your conscious perception of the sensory stimulus by electrically stimulating your sensory cortex. Therefore I canceled your perception in retrospect."

"You mean I apologized for an imaginary error. I reacted *properly* to a stimulus that I did not perceive?"

"Lesson 4: An experience at this very moment may be canceled later. Just as we learned in pure physics, there's also trouble with determining simultaneity in the brain. Moreover, the notion of cause and effect becomes difficult. The temporal order of subjective events is a product of the brain's interpretational processes, not a direct reflection of events making up those processes."

You turn to Constantia. "Next experiment. Constantia, look at this card. What do you see?"

"Two spots separated by about 4 degrees."

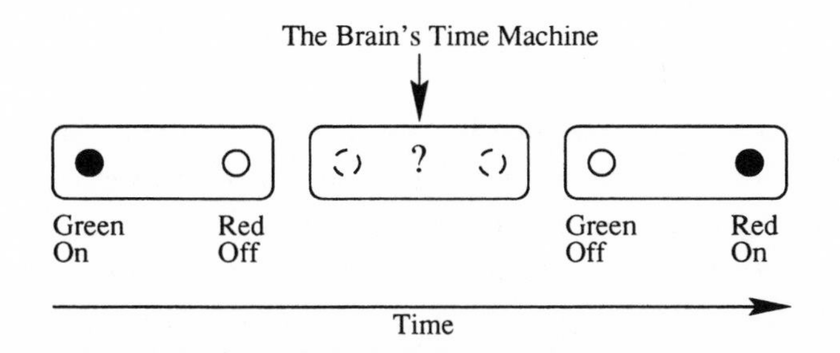

Figure 4.1 **A green light and red light are placed close to one another. The green light is flashed, and then the red light is flashed. What do you perceive in between the time the green and red light is flashed?**

"OK, I'm going to light them in rapid succession. What did you see?

"A single spot seems to move from one location to the other."

"What do you think will happen if I make one spot red and one spot green? What will happen to the color of the spot as it 'moves' from left to right?"

"The illusion of motion will disappear, replaced by two separate flashing spots?"

You turn to Mr. Veil. "Mr. Veil, what's your guess?"

"I think the illusory 'moving spot' will gradually change from one color to the other as it moves."

You turn on the device (Fig. 4.1). "Let's give it a try." A green light momentarily flashes on the left followed by a red light flashing on, on the left.

"Ooh," says Constantia.

"What do you see?"

"First the green spot seems to be moving, and then it changes to red abruptly in the middle of its (illusory) passage from left to right."

You nod. "Okay, you saw it smoothly traveling but abruptly change color. At the halfway point between lights, it turned from green to red. Here's my question for both of you. How were you able to fill in the spot at the times and positions between the two flashes of light, *before the second flash occurred*? How did the light 'know' to turn red before the red light came on?"

Mr. Veil comes closer. "Did we have clairvoyance that let us know the color of the second flash before it was experienced?"

You wave your hand in a modified kung fu outward block to dismiss Mr. Veil's idea. "The answer is that the imaginary content—the color change—couldn't have been created until *after* we saw the second colored light flash on."

Constantia takes a deep breath. "But if we were able to consciously identify the second spot, wouldn't it be too *late* to create the illusory color-switching and intermediate movement?"

You open a window to let in some fresh air. "Researchers have proposed that the intervening motion is produced retrospectively, but only after the second flash occurs, and is projected *backwards in time*."

Constantia brushes back some hair that a faint breeze has caressed out of place. "I like these perception experiments," she says. "Have you any other experiments we can do?"

"Certainly. Let's discuss one of my favorite temporal anomalies of consciousness. In the scientific literature, it's called the 'cutaneous rabbit.' Follow me."

You lead Mr. Veil and Constantia over to a velvet bench beside a doulcemele, a rectangular musical instrument with a keyboard. This is your favorite instrument in the museum; in 1360 it was presented to the King of France by King Edward III of England. Constantia sits down beside it and plays a phrase from Chopin's *Piano Concerto No. 2 in F Minor.* Even though the instrument has only thirty-five keys, her playing is quite good.

Just as she finishes her last note, you bend down. "Constantia, may I attach electrodes to your leg?"

Constantia considers the question for a moment. "Sure, I'll do anything in the name of science."

You place mechanical tappers on her ankle, knee, and thigh.

Constantia smirks. "Sir, couldn't this experiment have been done on my arm?"

"I suppose it could—"

"Then please do so."

"As you see best." You place the mechanical tappers on her wrist, elbow, and upper arm. "I will now deliver a series of quick rhythmical taps. Five taps on your wrist. Two at your elbow, and three more on your upper arm."

Constantia looks up from her arm. "It feels as if the taps travel in a regular sequence over equal-spaced locations on my arm—as if a little animal were smoothly hopping along the length of the arm from wrist to shoulder."

You nod. "Exactly. That's why it's called the 'cutaneous rabbit.' Cutaneous means 'of the skin.' Even though I made taps at only three places along your arm, you felt as if a rabbit were hopping *along* the arm."

Mr. Veil takes a close look at Constantia's arm. "Sir, how could Constantia's brain know that after five taps on her wrist there was going to be some taps on the elbow?"

"Neuropsychologists have determined that a person's experience of the 'departure' of the taps from the wrist begins with the second set of taps at the elbow. However, if we didn't deliver the elbow taps, Constantia would experience all five taps at her wrist, as expected."

"Sir, let me get this straight. Obviously, the brain can't know about a tap at the elbow until *after* it happens. Right?"

Constantia rubs her elbow. "Maybe the brain *delays* the conscious experience until after all the taps have been 'received,' and then it revises the sensory data to fit a theory of motion and sends this time-altered version of reality to our consciousness."

You sit down beside Constantia. "But would the brain always delay response to one tap in case one more came? If not, how does the brain know when to delay?" I don't know the answers." You pause. "Mr. Veil, would you like to be the subject of the next experiment? It deals with your mind playing tricks with time."

Mr. Veil executes a playful pirouette. "Yes!"

You walk over to Mr. Veil with an electrode in your hand. "Hold still. I'm now placing an electrode in your left cortex. I'm placing another on your left hand. I'm stimulating your cortex before your left hand is stimulated. Did you feel the tingles in this order?"

"No. I know that the left side of the brain controls sensation on the right side of the body. I'd expect to feel two tingles: first on my right hand, induced by the brain stimulation, and then on my left hand. What I did feel was reversed. First I felt the left hand tingle, then the right."

"It's all so crazy," Constantia said. "The timing of 'mental' and 'physical' events is all screwed up."

"Here's another experiment. Constantia, may I implant an electrode in your brain's motor cortex?"

She takes a step back. "Touching my leg is one thing. Sticking wires in my brain is another."

Mr. Veil grabs the wire from you and begins to insert it himself. "Sir, let me."

"Good. Now, I have wired you to a CD controller of Chopin's music. I want you to press a button on the CD to advance the music whenever you wish."

Mr. Veil begins the experiment, and the CD changes songs, but he seems confused. "Sir, the CD player seems to change songs before I have even decided to make a change! It seems to anticipate my decision."

"Quite right. But I tricked you. The CD button is a dummy button, and did not have any affect on the CD player. What actually advanced the songs was the amplified signal from the electrode in your motor cortex. Again we see that voluntary motions are not initiated by our conscious minds! The signal occurs before your conscious intent to push the button. Scientists called this experiment the *precognitive carousel* because they tried this test using a carousel and slide projector. Subjects advanced slides just as they were about to push the button!"

You remove the wire from Mr. Veil's head, and then you sit down on a piano bench. "Let's summarize. The mind plays tricks with time by making slight temporal adjustments. Today's lessons show that the brain plays with the concept of simultaneity and projects events back in time. You've also learned about cutaneous rabbits and precognitive carousels—all terminology in the serious psychological literature."

The three of you are silent for a while. You look at Constantia. "Constantia, would you like to have some dinner together? I could get reservations for two at the Four Seasons, or perhaps you'd prefer something more exotic, like the new sushi place, the Valldemosa."

Constantia pauses for a few seconds considering your question. "I think that would be nice. But Mr. Veil—"

Mr. Veil waves his forelimbs. "Oh don't worry about me. We don't enjoy the food you Earthlings enjoy. Frankly, I prefer something more jumentous—"

Before Mr. Veil finishes, you take Constantia's hand and smile at her. "Perhaps while you are eating, you wouldn't mind me placing electrodes on your scalp. Won't hurt a bit. You see, I'd like to study—"

Before you finish your sentence, Constantia throws the CD player at your head. It ricochets off a nearby harpsicord producing a sound like the chanting of metallic monks. ✍

The Science Behind the Science Fiction

> Science fiction writing, at its finest, is an enormous leap of skeptical imagination. The art is in choosing which impossibilities to shatter, which future advancements to promote to reality.
>
> —Greg Bear, *Skeptical Inquirer,* 1996

Time and Consciousness

The experiments in this chapter demonstrate how the mind plays tricks with time, and are all based on numerous different physiological experiments conducted by Wilder Penfield, Benjamin Libet, Paul Kolers, M. von Grünau, H. van der Waals, C. Roelofs, and others, and described by Daniel Dennett, Marcel Kinsbourne, and Rainer Wolf (see References). Voluntary motions are not initiated by our *conscious* minds. And the brain (Fig. 4.2) does seem to have a "time machine" for antedating perceptions. The brain "projects" mental events backwards in time in strange ways. In fact, serious chronopsychologists use physical-sounding terms such as "horizon of simultaneity" (see Pöppel references in Dennett) when referring to the brain's time machine.[1]

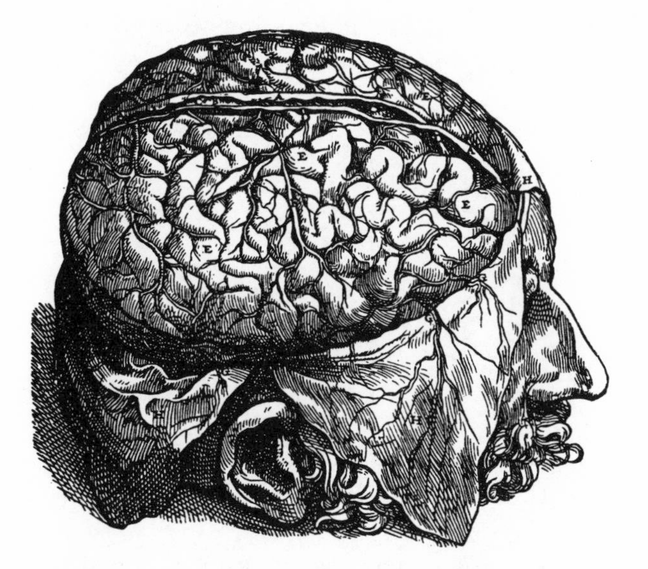

Figure 4.2 The brain is a time machine. Voluntary motions are not initiated by our *conscious* minds. And the brain does seem to have a "time machine" for antedating perceptions.

Figure 4.1 is a schematic illustration of the green/red light experiment discussed in this chapter. Subjects actually report seeing the color of a "moving" spot switch midtrajectory from green to red. How are we able to fill in the spot at the intervening places and times (along a path running from the first to the second flash) *before* that second flash occurs? One possible mechanism is that the brain has an editing room located prior to consciousness. Perhaps there is a time delay, as is common in "live" TV or radio broadcasts, that allows time for the brain to censor stimuli before consciousness. In the brain's editing room, in-between frames are filled in after the red light flashes, and the brain inserts these back in perceived time like a director splicing in new frames in a videotape. By the time the finished piece arrives at consciousness, it already has its illusory insertion.

Figure 4.3 reemphasizes what you have learned in this chapter, namely, that the time line experienced by your conscious brain is often quite different from the "objective" time line of events occurring in your brain or in real life. In short, the time lines do not register and can develop kinks and order differences relative to another. Daniel Dennett and Marcel Kinsbourne believe that this misalignment is "no more mysterious or contra-causal than the realization that the individual scenes in movies are often shot out of sequence." Nevertheless, the misalignment of time lines should stimulate fascinating adventures and experiments in consciousness, particularly as research in computer–brain interfaces matures in the twenty-first century.

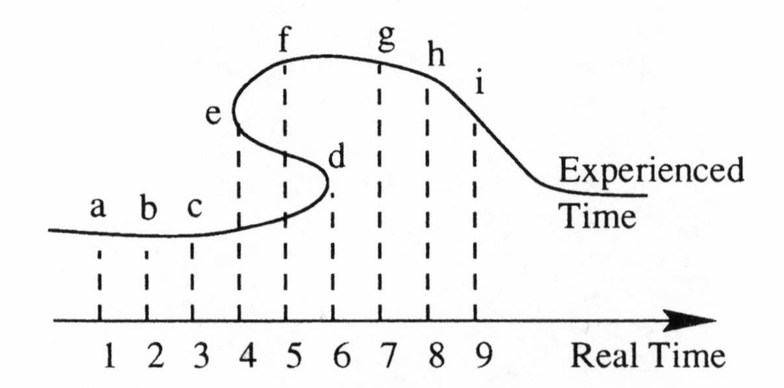

Figure 4.3 Time-line kinks. A subjective sequence of experienced events (top line) often does not coincide with events in the brain (bottom line). For reasons discussed, order differences in the experienced time (top line) may cause kinks in its time line relative to events in the brain.

Researchers have also conducted experiments relating to free will. For example, a subject, with brain waves monitored, is asked to flex his wrist whenever he feels like it. The subject watches a rapidly moving clock and reports the exact time of his subjective intention to flex his wrist. The electrode responds about 300 milliseconds before the decision is experienced. (This means that you can know a person's intent before the person does!) The sensation of free will, created in memory, occurs after the fact as a convenient way to record the decision.

Other time experiments involve alternately flashing police car lights that can appear to move back and forth. Moreover, one light appears to move toward the other a split-second before the other flashes. Obviously the brain cannot predict the future and know the other light is going to flash before it does. The whole experience of movement is created after the fact. All experiences are created after the fact.

Readers interested in various temporal anomalies of consciousness can read Dennett and Kinsbourne's article, "Time and the Observer."

The next few sections include quick "what-if" exercises in the psychology of time.

Sloth and Hyperland

If you are a teacher, have your classroom imagine an island in the Pacific, or a peninsula off America, where there is something in the air that induces an altered state of consciousness in which time is perceived to pass at a slower or faster rate than normal. Call this slow land Sloth. In Sloth, you would sleep for days. It would take fifteen minutes (of normal Earth time) simply to sneeze. On the other hand, consider Hyperland where an automobile might be constructed in seconds, or a book written in an hour. How would these societies interact with the rest of the world? What would be the benefits of living in either land? What would happen to the world's economic trade and geopolitics?

Drugs, Babies, and Time

Do newborn babies have a sense of the passage of time? Philosophers like Immanuel Kant thought so, and they believed that time was something we experience directly from birth, that it exists outside of us. Other philosophers believe that time is a construct of the human mind. For example, philosopher Henri Bergson treats time as something derived entirely from subjective experience. According to Bergson, an infant would not experience time directly but rather have to learn how to experience it. If time is something learned, can we unlearn it?

When Einstein was once asked about "psychological time," he remarked, "When you spend two hours with a nice girl, you think it's only a minute. But when you sit on a hot stove for a minute, you think it's two hours." Since Einstein, there has been increasing research on the psychology of time dilation. For example, sleep studies show that, during dreaming, time is dilated: During brief periods of (external) time, there can be long sequences of internal events. For example, if I woke you after you dreamed for five minutes, you could tell me a long sequence of events that appeared to have taken much longer than five minutes.

Your psychological perception of time is, of course, affected by such things as medications, time of day, your level of happiness, external stimuli, and even the temperature. Hypnosis can also cause time dilation, as can cannabis and LSD. Additionally, heat appears to speed up the activity of a chemical timepiece in the brain. For example, fever can severely speed your perception of time, perhaps partly because it speeds chemical processes. Opium is notorious for its effect on time perception. The English writer Thomas De Quincey reported that under the influence of opium he seemed to live as much as one hundred years in a single night. Another Englishman, J. Redwood Anderson, took hashish and said, "Time was so immensely lengthened that it practically ceased to exist." (This reminds me of Tennyson's Lotus Land "where it was always afternoon.")

Even without drugs people can learn to stare at the second hand of a clock and perceive it to stick, slow down, and hover. This takes training, but some people can experience the hand to stop altogether for a while. Some psychologists propose that the observing mind, the entity that correlates and makes sense of information submitted to it by the brain, is temporarily absent during these time-sticking periods. The brain hardware is left unattended while the mind has gone elsewhere.

Interestingly, a person under hypnosis can judge time more accurately than he can when in a normal waking state. For example, if a hypnotized person is asked to awake after five minutes, he can judge this time interval more accurately than normal. This leads me to believe that unconscious perception of time can be more accurate than conscious ones. By this I mean that the brain can be "trained" to measure certain time intervals that the conscious mind cannot measure. If I make a buzzing sound and then nine seconds later flash a light in your eyes, and do this over and over, your brain will eventually show a conditioned reflex: After I make a buzzing sound, nine seconds later your brain wave will change. However, if you were asked to make a conscious estimate of the delay between sound and light your guess would be much less accurate.

Mental Illness and Time

Autistic savants have serious mental handicaps but also sometimes have spectacular islands of ability or brilliance that stand in stark contrast to their extremely low IQs. Savants seem to possess brain functions that are beyond our complete understanding. (Some of you may recall the cinematic portrayal of an autistic savant in the 1988 movie *Rain Man,* which starred Tom Cruise and Dustin Hoffman. In that film, Hoffman played Raymond, a middle-aged autistic who displayed many of the abilities of the savant including counting, memorization, and calculations.) Some savants have an incredibly developed sense of the passage of time. For example, one savant in the literature could tell, to the minute, the exact time at any time of the day or night, but could not read a clock. Another knew exactly when commercials would begin and end, even when out of range of the TV. Some savants can tell exactly how much time has passed during a specific period without looking at a watch or clock.

In contrast to some autistic savants, schizophrenics are often unable to cope with time. In the words of one patient, "I just can't seem to grasp the fact that time passes and the hands of the clock go round. Sometimes, outside in the garden when they run quickly up and down . . . or the leaves whirl in the wind, I wish I could live again as before and be able to run with them within me so that time would pass again. But there I stop and I do not care . . . I just bump into time."

Your Life on a Paper Strip

In a past study, when subjects were handed strips of paper and asked to mark different periods of their lives, their accuracy fluctuated markedly with their ages. For example, all tested subjects emphasized recent events like "yesterday" and "last week" too strongly by placing them farther to the left than they should be. Of the test subjects used for Figure 4.4, the seventy-year-old man was the most accurate in properly segmenting the five time periods spanning his life. Notice that the nine-year-old allocated the same length of strip to last week and six months ago. Time sense appears to get better with age.

These and other anomalies have led my readers to propose unusual ways of representing lifespans on paper. For example, Jeremy Weinstein of Walnut Creek, California, wrote to me suggesting experiments with

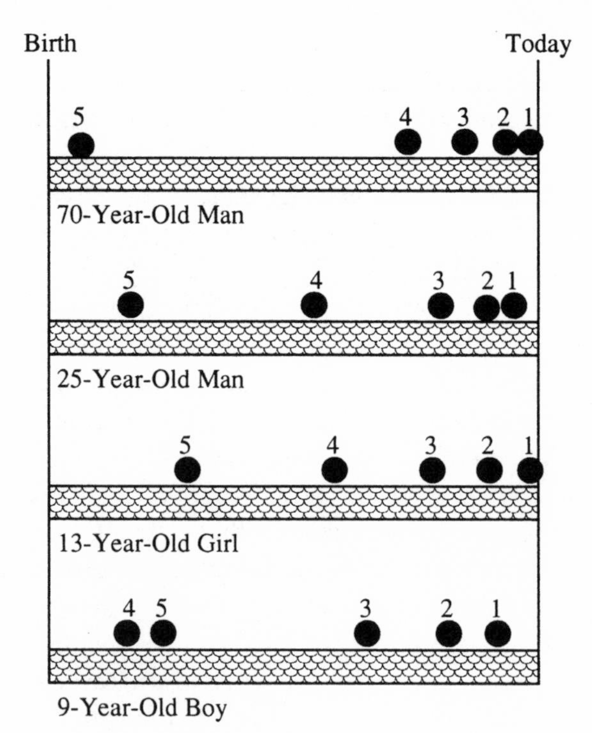

Figure 4.4 Subjective time. Experiments reveal that the ability to assign a correct time relationship between past events varies widely with age.

"human life-span perception plots." In particular, he plots a human life-span in terms of its "perceived length." In other words, the first year of life is denoted by "1". The second year is 1/2 (because, to the two-year-old, his second year is half of his life), the third year is another third, and so on quasi-logarithmically until the seventieth year is 1/70th. Weinstein remarks, "Using this method, an 82-year-old's life is not half over at 41 but rather at 7."

Romantic reality is the cutting edge of experiences. It's the leading edge of the train of knowledge that keeps the whole train on track. The leading edge is where absolutely all the action is. The leading edge contains all the infinite possibilities of the future. It contains all the history of the past. Where else could they be contained?"

—Robert Pirsig, *Zen and the Art of Motorcycle Maintenance*

168

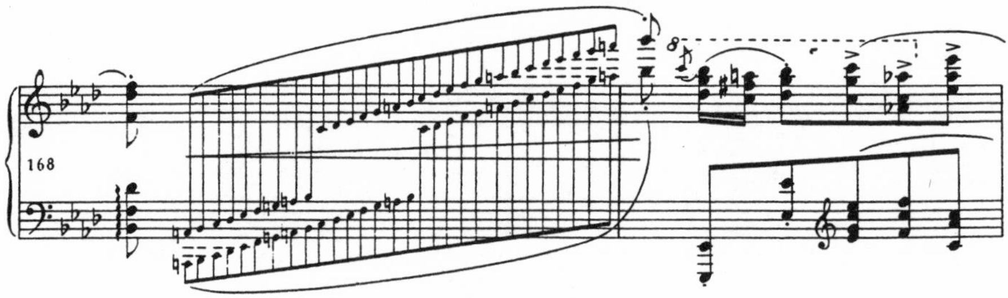

Everything used to measure time really measures space.

—French philosopher Jerome Deshusses

The past cannot remember the past. The future can't generate the future. The cutting edge of this instant right here and now is always nothing less than the totality of every thing there is.

—Robert Pirsig, *Zen and the Art of Motorcycle Maintenance*

Here–Now and Elsewhere in Spacetime

5

✍ "Sir, snap out of it."

"What?" You take a deep breath. Very faintly, off to your right, you can hear many clocks ticking softly.

Mr. Veil's large eyes are gleaming. "We're ready for our next lesson."

You have been daydreaming about a cottage on a prehistoric beach—a retreat for you and Constantia back in time. In the cottage is Chopin's very own piano, borrowed from the 1800s, and you are playing his *Piano Concerto No. 1 in E Minor* as Constantia reclines on a sofa made of triceratops hide. High in the sky, in the shape of a "V," a flock of pterodactyls glides by with wings large enough to blot out the sun.

"Sir?" Mr. Veil says.

You shake your head and look from Mr. Veil to Constantia. Her shirt swims with repetitions of the formula "$E = mc^2$" in various-sized fonts. Her shirt's buttons are caricatures of Albert Einstein.

You are standing in Contantia's back room (and living quarters) in the museum. Mr.

Veil motions you in the direction of three chairs he has brought in from his own sleeping quarters. You sit down. The chairs are fairly comfortable, considering they were designed for a race with segmented rear ends that looked like cracked triangular tortilla chips.

Constantia has chosen a large room that served as a research library and lab, and also as a sleeping quarters. You look around and feel dwarfed by the soaring pilasters, frescoed ceilings, and gilt furniture. On the wall is a photo of Constantia and a handsome man, both playing cellos before a large audience in St. Petersberg. There is an inscription on the photo in a tiny, almost childish hand. You see only Constantia's face, but her eyes are unsettling. What are her feelings for the man? You feel like removing the picture from the wall.

On a workbench by her bed are half a dozen dismantled Stradivariuses with their bellies and backs attached to vices. Constantia raps on one of the backs. "I've got to find out what made the Stradivarius tick," she says. "The Cremona-style violin makers soaked the wood in seawater. That's one of the reasons the violins sounded so wonderful."

Constantia's fingers trail lingeringly over the back of a violin, and then up its thin neck. They travel back and forth along the ancient brown wood. Occasionally she plucks a string, setting the violin all aquiver.

You shake your head. "Before we discuss spacetime diagrams, I'd like you both to make a list of ten events in history you'd like to go back and watch—aside from Chopin's playing."

Constantia grabs a paper and pencil. "What a wonderful exercise," she says.

After a few minutes, Mr. Veil and Constantia say, "Ready."

"Here's what I'd like to watch," you say. Your lists reads:

- Moses leading the Jews from Egypt, or receiving the Ten Commandments
- The crucifixion of Jesus
- Baha'u'llah in prison
- Muhammad fleeing from Mecca
- Any hypothetical UFO abductions
- Dinosaurs walking the Earth

- Sermon on the Mount
- Chopin performing the *Scherzo in B Minor.*
- The eruption of Vesuvius and the death of Pompeii
- Cleopatra making love with her famous paramours
- O.J. and Nicole Simpson, on the night of her murder

Constantia's list reads:

- Athen's democratic government
- Manifesto of Marx and Engels launching communist movement
- Charlemagne taking the battered crown of the Roman Empire
- The fall of the Eastern empire to the Turks
- The battle of Poltava where Peter the Great defeats the Swedish
- Pablo Picasso painting *Guernica*
- Strauss waltzes, conducted by Strauss himself
- The beheading of Ann Boleyn
- The disastrous premier of Stravinsky's *Rite of Spring*
- Napoleon at Waterloo

Mr. Veil's list reads:

- Euclid establishing systematic geometry
- Newton's *Principia* jolting scientific world
- First black men sold in Jamestown, Virginia
- Black slavery outlawed by emancipation edict
- Beethoven's symphonies, conducted by Beethoven himself
- Plato teaching Aristotle
- The burning of the Hindenburg zeppelin
- Kennedy's assassination
- Elvis's last concert

Constantia's eyes suddenly seem to glow with a savage inner fire. "Sir, I have another to add to my list. I'd like to go back in time and watch the eighteenth-century violin makers. I still can't figure out how the semiliterate craftspeople of Cremona were able to achieve what has eluded everyone else with better education."

"Excellent," you say as you place your hand on hers. "If we could decode the Stradivarius, we could solve one of the greatest mysteries of science and art, not to mention making a million bucks."

Constantia removes your hand from hers in about five seconds. (At least this is better than her one second removal-time of yesterday.) Perhaps there is still hope.

You begin to pace. "You can tell a lot about people by the lists they make."

Constantia puts her hands on her hips. "Yes, why was Cleopatra on your list?"

You wave your hand. "But let's put our lists away for now. Who knows. We might have time to do a few of these after we visit Chopin."

You walk over to a small blackboard hanging on the wall. "Today I want to teach you about spacetime diagrams, light cones, world lines, superluminal motion, geodesics, and memories. Tomorrow, let's make some time to study distance metrics, imaginary time, and Minkowskian space. Let's take it slowly. I'd like to talk in some depth about things that will be important to us for time travel."

Constantia's mouth hangs open, not in horror or despair but rather in breathless anticipation and delight, or so it seems. Mr. Veil's tail whips are all aquiver. It's wonderful to have such enthusiastic and interested students. But how could it be otherwise? You are a great lecturer, and this is all a prelude to your actual adventures back in time. The physics are a means to this end.

"Let's start by making *spacetime diagrams.* These will help us understand the physics and use a time machine. These kinds of drawings show our position in space and time. The resulting curves are our *world lines* showing our paths through space and time."

Mr. Veil raises his hand. "Sir, there are three space dimensions and one time dimension. Wouldn't these diagrams be hard to draw?"

"We're going to simplify them. Let's start by using just one dimension of space. The space axis in our two-dimensional plot is the x-axis. The time dimension is the y-axis."

You take out a piece of chalk and begin to sketch.

"Who wants to be in our spacetime diagram?"

Mr. Veil jumps up from his chair and bangs into Constantia. "Me,"

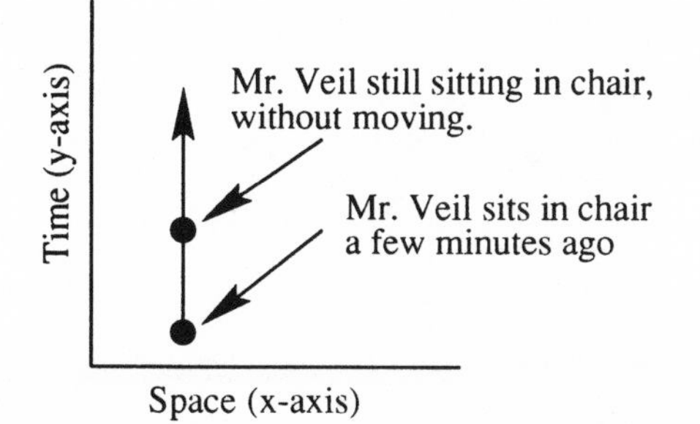

Figure 5.1 World line of a stationary object. The value of its space coordinate never changes, as time proceeds "up the page."

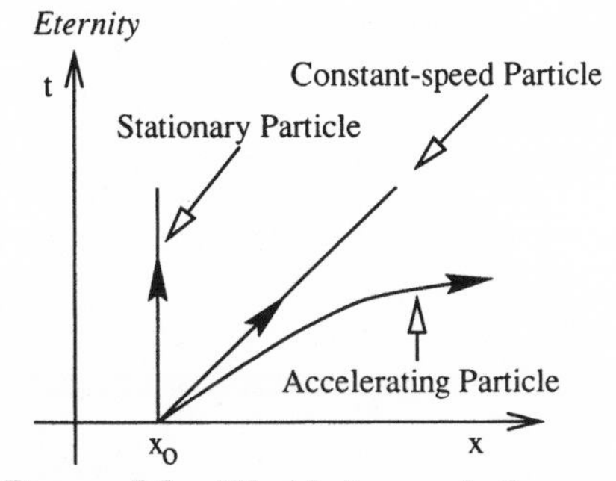

Figure 5.2 World lines of three objects through spacetime.

he shouts.

"Will you watch out?" Constantia yells at Mr. Veil.

"Fine," you say. "Sit down, Mr. Veil. Let's start by assuming you're not moving, and draw a spacetime diagram for you." You draw a straight line on the board. "Because you are never moving, your x-value doesn't change, but your time-value progresses in the direction of the arrow. You're world line is straight" (see Fig. 5.1).

Now you give Mr. Veil's chair a shove. "If you were to move at a constant velocity, your world line would tilt away from the vertical, and if you were to accelerate, your world line *curves* away from the vertical" (see Fig 5.2).

Constantia examines the diagrams. "Sir, you mean that straight, uncurved, world lines are for unaccelerated bodies experiencing no forces?"

"Correct. And in my diagram, I assumed that Mr. Veil started sitting in his chart positioned at x_0 when $t = 0$. Straight world lines are called spacetime *geodesics*."

"Now let's talk about *light cones* and how they represent the fabric of reality. Traditionally and for convenience, on spacetime diagrams we assume the speed of light c is equal to one. This means that a distance of 300,000 kilometers on the x-axis corresponds with one second on the t-axis, and we make the intervals along x and y equal for these two values. Can either of you draw the world line of a photon?"

Mr. Veil gets up at the same time as Constantia, who promptly pushes him back into his seat. She goes over to the chalkboard. "I think I can handle this," she says. "The world line of a photon, which travels at the speed of light, is tilted away from the vertical time axis by 45

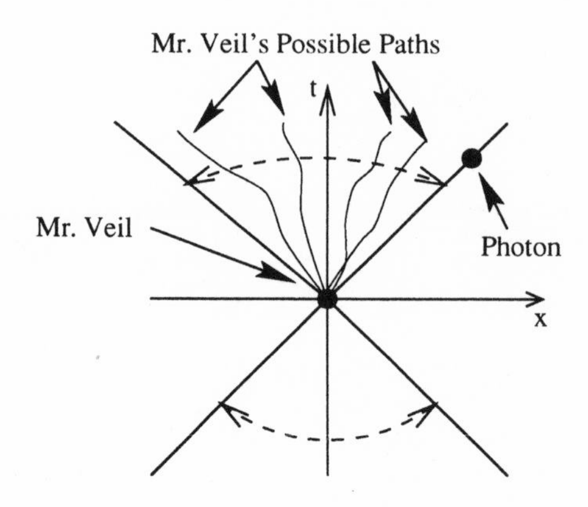

Figure 5.3 World lines confined to sublight speeds that never tilt more than 45 degrees from the vertical.

degrees" (see Fig. 5.3.)

"Yes! And because photons can travel in both space directions in our diagram, and the speed of light appears to be a limiting speed, all world lines must be paths that never tilt more than $\pm$ 45 degrees from the vertical." You pause. "Now, let's add an extra dimension of space to the diagram so that the straight world lines of the photon trace out a cone. Inside the cones are all possible world lines involving speeds less

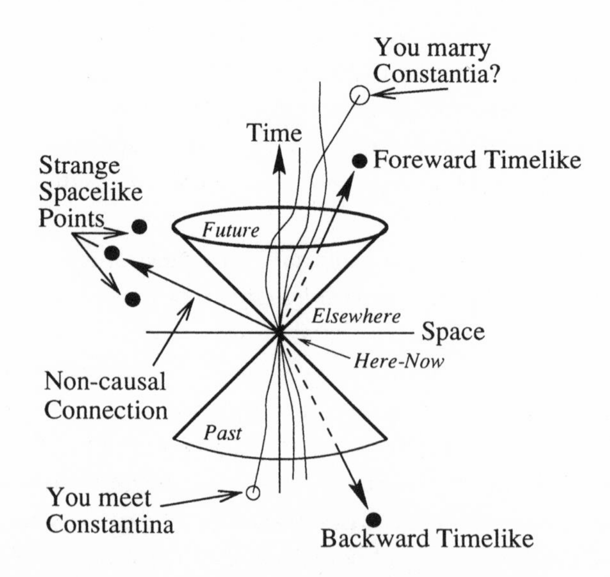

Figure 5.4 A light cone with spacelike and timelike world lines.

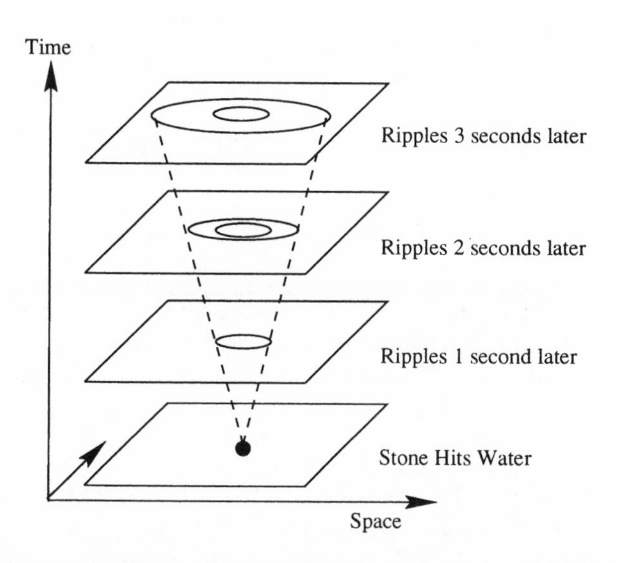

Figure 5.5 Circular ripples on a pond spread out in time and define a cone reminiscent of light-cones in a spacetime diagram.

than the speed of light for Mr. Veil located at $x = 0$ and $t = 0$. This location in spacetime is called the *here–now*" (see Fig. 5.4).

Constantia sits down and raises her hand. "Sir, all the world lines in the upper cone are in the future for Mr. Veil in the here–now. All points in the bottom cone are in Mr. Veil's possible past."

"That's right. The regions within the cones are called *light cones.* We can draw a straight world line from here–now to a point in the *future cone* with a tilt of less than 45 degrees away from the vertical. That means a particle could travel from Here–Now to the point at a speed less than the speed of light. These world lines inside the cone are called *timelike* because their projection on the time axis is greater than their projection on the space axis. These world lines have the possibility of causally linking events."

Constantia gets up and begins to draw on the light cones. She places herself at the center and draws a dot on a past world line and on a future word line. "Let's see if I understand. This bottom dot in the past cone represents our first meeting in the museum. Pretend we got married at the here–now point. This upper dot in the future cone represents the time and place we would have children. The world line is continuous from past to future and is contained within the double

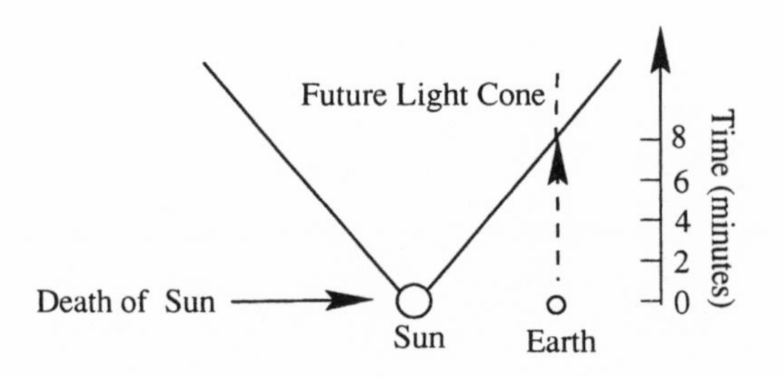

Figure 5.6 In the "elsewhere." If the sun were suddenly destroyed, we would continue to see light from the sun for about eight minutes, because it takes light eight minutes to reach us from the sun. We on earth would initially be in the elsewhere of the destruction event.

cone. Our meeting in the past is an event, or cause, which can have an effect on the here–now because its influence propagated at less than the speed of light. Also, a cause at here–now can affect an event at any point in the future cone."

"Constantia, d-does that mean we're getting married?"

She smirks. "We're talking hypothetically."

You draw another diagram (Fig. 5.5). "Another good way to visualize light cones is to imagine a stone dropping in a pond. The ripples spread out on the surface of the pond as time progresses. If we imagine a three-dimensional model consisting of a two-dimensional pond surface and one dimension of time, the expanding circular ripple encompasses a cone whose tip is at the time and place at which the stone hit the water. In the same way, light spreading out from an event forms a three-dimensional light cone in the four-dimensional space time."

Mr. Veil begins to stare at your previous diagram. "Sir, what do points outside the future and past cones mean?"

"They can't be reached from here–now except by world lines tilted more than 45 degrees away from the vertical. If we draw world lines to reach these points we would call them *spacelike* because the lines' projection onto the space axis is greater than their projection on the time axis. The lines represent travel at speeds greater than light. It's impossible for these world lines to causally connect events. The set of all world lines outside of the light cones form the *elsewhere of here–now*. Let's consider a concrete example. If the sun were suddenly destroyed by an alien race, we would continue to see light from the sun for about eight minutes, because it takes light eight minutes to reach us from the sun.

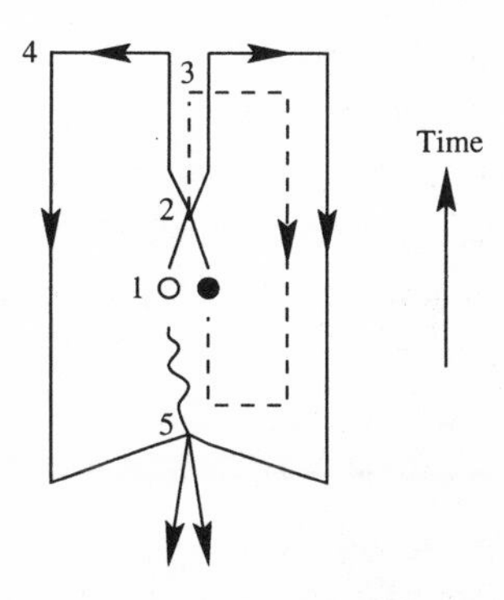

Figure 5.7 If time travel is possible, then world lines might become closed loops. Here you meet Constantia (1), and have a baby daughter, Constantia Junior (2), represented by the dotted line, who grows up (3) and decides to travel back in time. Constantia junior grows up and meets you at "1"! Meanwhile, at the position marked "4" you and Constantia Senior decide to go back in time. At "5" you have a baby boy who grows up (wiggly line) to be you at "1." At the bottom of the figure, you and Constantia go back and visit the dinosaurs. Notice that Constantia is her own mother and grandmother, and you are your own father and grandfather.

We on earth would initially be in the elsewhere of the destruction event. Only after eight minutes would events on Earth lie in the future light cone of the event at which the sun blacked out" (see Fig. 5.6).

Constantia caresses back a lock of hair that has fallen onto her cheek. "Does every point in spacetime have its own light cone?"

"Yes, if event 2 is in the future cone of event 1, then event 1 must be in the past cone of event 2."

Mr. Veil extends his left forearm toward you. "Sir, why are we doing all these exercises? If it's simply to go back to Chopin's time, wouldn't it be so much easier to watch someone else play his music on the television?"

You shake your head. "TV alienates me. When we actually see Chopin, back in time, there's no TV frame to distract us. We'll be completely in contact with it all. We'll be living it. We'll be in the scene, not just watching it. The sense of presence, the whispers of the audience, the sight of his flying fingers, should be overwhelming. His fingers will move so fast, you'll swear each had a little time machine strapped onto its knuckle. Look at his musical scores. How could humans play that? His music, more than anyone else's, seems to violate causality."

Constantia starts plucking the strings of a balalaika, a Russian musical instrument of the lute family. "What's this I keep hearing about 'causality violation' in time travel?"

You listen for a while as she masterfully manipulates the three strings. The instrument was usually used in folk music or in large balalaika orchestras in the twentieth century. Constantia, however, gave the instrument a New Age sound. You return your thoughts to time travel. "Traditionally, scientists had thought poorly of anyone who believed in the possibility of time travel. Causality—the notion that every effect is preceded, not followed, by a cause—is firmly entrenched in the foundations of modern science. But there are many modern physical theories, for example time travel by using wormholes, that don't seem to violate physical laws but do violate causality. We'll talk about this some more another day." You pause. "Let me give you examples of time-travel paradoxes involving causal loops. I'll draw lines to represent our world lines, our paths through space and time."

You draw a rough, schematic diagram (Fig. 5.7). "Now this is a wild example. In this figure, I represent myself as the open circle in the center and Constantia by the filled circle. Let's assume we initially meet at the position in spacetime marked by the '1.' A little later, at the position marked by the '2,' we marry and have a baby daughter, Constantia Junior. Her path through life is represented by the dashed line. Unfortunately, Constantia Junior is abducted by a stranger at birth, and we never see her again. She grows up, and at age twenty (marked by '3') she decides to go back in time and find her roots. After traveling back in time, she spends twenty years growing up and having a fairly normal life. Finally, she meets me at '1'! We fall in love, marry, and the rest is, as they say, history. She is the woman I initially met at '1.' Meanwhile, at the position marked '4', the 'original' Constantia senior and I decide to go back in time in hopes of finding our lost daughter. We go back in time and at '5' we have a baby boy who grows up (wiggly line) to be me '1.' At the very bottom of the figure, the 'original' Constantia and I go back and visit the dinosaurs. Notice that Constantia is her own mother and grandmother, and I am my own father and grandfather."

Constantia stops playing the balalaika. "Crazy!" she says.

You nod. "Who is Constantia's mother, father, grandfather, grandmother, son, daughter, granddaughter, and grandson? Constantia Junior and Senior are the same person. If we draw more of Constantia's family tree, we might find all the branches are curled inward back on themselves, as in a loop. She can be an entire family tree unto her-

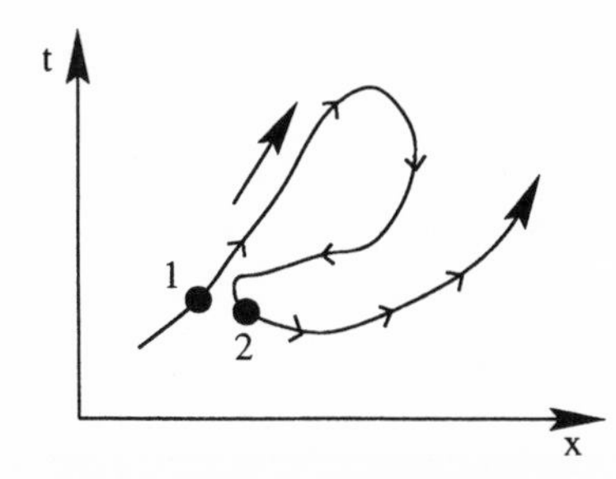

Figure 5.8 World line of a person traveling backward in time from 1 to 2.

self. This is an example of a paradox unlike the one where a person goes back in time and kills his grandmother, thus altering the past. In the case for which I drew a diagram, you are *fulfilling* the past, not destroying it. Thus the world lines travel in a *closed loop,* fulfilling rather than changing the past."

Constantia sits by the balalaika and straightens out her legs, lifting one at a time slowly without looking up. You've noticed that long silences mean sadness for her.

"Everything okay?" you say.

She looks up and then down again. "It's hard to believe that we're really going to travel back in time? These lessons are great, but—"

"We will go back, but I want to give you the lessons first. If we were to suddenly jump into a time machine and go back in time right now, you'd see Chopin and time travel in one kind of context, almost like an impressive magician's trick. But to arrive after days of hard lessons would be to see him in another way, as a goal, a promised land, a completion of understanding. That's why I like teaching. That's why I don't simply want to use technology, but I want to understand it."

Constantia begins to smile. "Okay. We want to go back in time. What would our world lines look like? Does a spacetime diagram always need to have future-directed world lines?"

"If a particle could move back in time, the diagram can show this by having the world line switch direction and curve back on itself. It can come arbitrarily close to itself."

You sketch on the board (Fig. 5.8). "I've drawn the world line for Mr. Veil traveling backward in time from Event 1 to Event 2. In fact, here Mr. Veil is visiting himself in the past!"

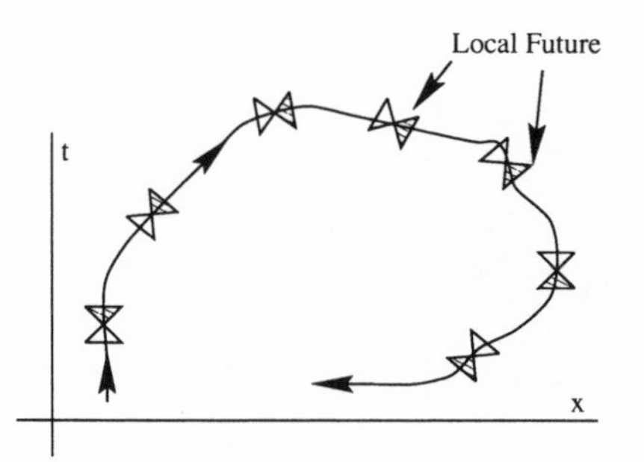

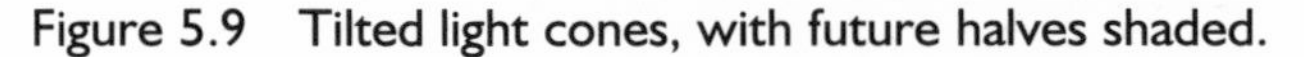

Figure 5.9 Tilted light cones, with future halves shaded.

Mr. Veil takes a piece of chalk and draws on the board. "Sir, what if the line loops back and touches itself near 1 and 2?"

"Mr. Veil, that would be viewed by some physicists to be catastrophic. In my drawing, the world line did not touch or cross itself. In your diagram, you would be not only be visiting yourself in the past but you would be occupying the same spatial location at the same time as your earlier self."

Constantia comes closer. "Wouldn't that be impossible—because since it did not happen, then it cannot happen."

"Constantia, time travel can create a lot of paradoxes. Some believe that to avoid the paradoxes, a new universe would be created each time one went back in time. In these universes, the future is different from our own universe."

You point to the figure again. "Notice that the little arrows on Mr. Veil's world line are always in the direction of the *local* future of Mr. Veil. His memories are formed in the direction of the arrows. He'd have more memories at location 2 than at location 1 even though the two spacetime positions are almost identical."

Mr. Veil jumps and bangs into Constantia again. "But wait, Sir. There's a problem. How can we draw a world line that loops back in such a way that it never tilts more than 45 degrees at all places? At some point, it must be bent more than 45 degrees."

"You're right, at some point along a backward time-traveling world line, $|dx/dt| > 1$. This would represent *superluminal motion*—faster than light travel. The term *dx/dt* is the slope of the line."

"Oh, no!" Constantia says. "How will we ever go back in time?"

"Don't worry. One way we can allow a subluminal world line to loop back is to arrange a string of light cones on the world line in various tilted orientations." You sketch on the board (Fig. 5.9).

Constantia stares at the diagram. "Is—is that possible?"

"Yes, if spacetime is curved. In fact, we're going to use a Tipler cylinder to curve spacetime; in this way the light cones will be tipped, and each future cone points into the past cone of the next here–now. We'll talk more about that soon. First we need to study more of the preliminaries."

Mr. Veil's strangely articulated legs move in an oddball synchrony, giving him the appearance of a drunken spider. "Sir, what happens if I remained stationary in space. Could time travel be possible? Or do I have to move?"

"Time travel would be difficult if you don't change your location in space, because your world line would have to bang into itself. Try drawing the world line as it reverses its time direction. That would be a real mess. The only way the world line can bend back on itself is for you to change your position in both time and space. Some science-fiction authors seem to get that wrong. Their characters enter a time machine that stays still, and then they climb out. If we really want to deeply analyze science-fiction stories, we'd have to consider that the Earth is moving, and this should cause the time travelers to materialize somewhere in outer space rather than where their time machine originally was. In fact, the Earth moves though space in a very complex way. A time machine sitting in a living room is moving about the Earth's axis. The Earth is moving around the center of gravity of the Earth–Moon system, and also about the center of gravity of the Earth–Sun system. The Earth also accompanies the Sun in its motion around the center of the galaxy, and the galaxy moves relative to the center of gravity of the Local Group. The Earth moves around the galactic center at speeds close to 200 kilometers per second."

You look out a window of the museum. The pigeons are back, and everything seems so alive. The people, the birds, even the cars. Every foot of Fifth Avenue is humming with action, a whole community of millions of organisms going through life in a kind of benign continuum. Perhaps you are spending too much time in the stuffy museum? You need fresh air, and yearn for travel. You look out again and watch as dusk fills the roads with purple mist, and faint puffs of vapor hang over the sodden cement.

You pick up another balalaika in Constantia's room and begin to play. Constantia accompanies you with her own instrument. As the music evolves, Mr. Veil begins to gyrate, his whip-tails intertwining like stripes on a candy cane. Shadows begin to spring up in the room as if they are living creatures. Soon the only illumination comes from green and red lights emitted by bioluminescent bacteria coating Mr. Veil's exoskeleton. It reminds you of Christmas. ✍

The Science Behind the Science Fiction

Taken to its ultimate, the cumulative audience paradox yields us the picture of an audience of billions of time-travellers piled up in the past to witness the Crucifixion, filling all the holy land and spreading out into Turkey, into Arabia, even to India and Iran. . . . Yet at the original occurrence of that event, no such hordes were present. . . . A time is coming when we will throng the past to the choking point. We will fill all our yesterdays with ourselves and crowd out our own ancestors.

—Robert Silverberg, *Up the Line*

In all time-travel stories where someone enters the past, the past is necessarily altered. The only way the logical contradictions created by such a premise can be resolved is by positing a Universe that splits into separate branches the instant the past is entered.

—Martin Gardner, *Scientific American,* 1979

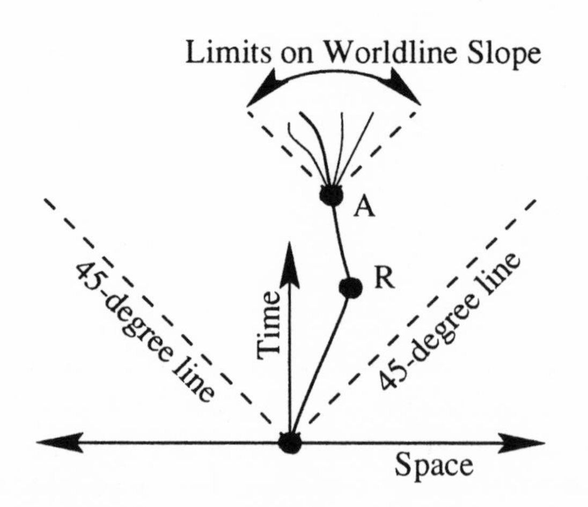

Figure 5.10 Constraints on world lines.

World Lines

World lines can be used to visualize the limits that the universe seems to place on motion. For example, in Figure 5.10 we see various world lines available to an object, such as a spaceship, at position *A*. Because an object must move at sub-light speeds, its future world line is confined to a region delimited by 45-degree lines on a spacetime plot. These limits on the slope apply to every point on the object's world line. For simplicity, the units of space and time can be measured in the same units and plotted to the same scale along the graphs' axes. (Recall that time units, e.g., seconds, can be converted to distance units, e.g., meters, by multiplying the time values by the "conversion factor" *c*, the speed of light.)

A world line gives a complete description of an object's motion in spacetime. By examining a the world line, we can tell an object's position and velocity at every moment along its path. For example, we can tell that the spaceship is at rest at position *R* because the world line has a vertical tangent at this point. In contrast, we cannot tell an object's speed by looking at a picture of its *trajectory* or *orbit* in space.

If you are a teacher, have your students draw world lines for various situations—meeting their grandfathers, going to the future and visiting creatures on a nearby star, freezing time and wandering through a crowded Manhattan street, meeting a person living forever, and so on. Also, have your students make a list of their favorite places to visit in history.

Our world lines never really begin or end. When we die, the world lines of the molecules in our body keep going. When we are born, the world lines of molecules from our mothers coalesce into our embryonic forms. At no point do world lines break off or appear from nothing. According to Einstein, world lines cannot be cut. This means if we went back in time and killed our mothers before our births, we would not disappear as depicted in such movies as *Back to the Future.* Thus, altering the past is not possible using Einstein's relativistic ideas.

As we age, the molecules in our bodies are constantly being exchanged with our environment. With every breath, we inhale the world lines of hundreds of millions of atoms of air exhaled yesterday by someone on the other side of the planet. In some sense, our brains and organs are vanishing into thin air, the cells being replaced as quickly as they are destroyed. The entire skin replaces itself every month. Our stomach linings replace themselves every five days. We are always in flux. A year from now, 98 percent of the atoms in our bodies will have been replaced with new ones. We are nothing

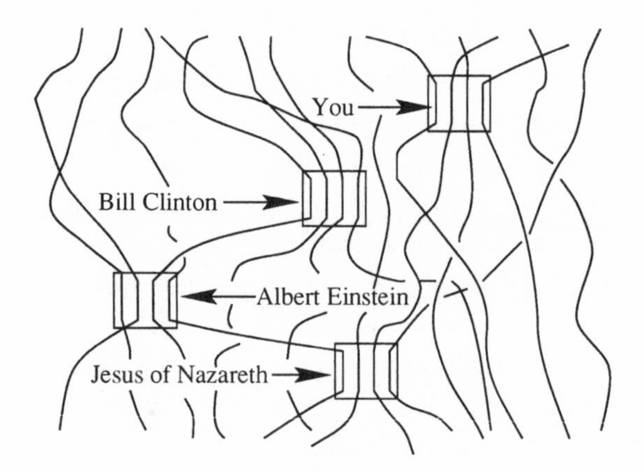

Figure 5.11 People are spacetime tangles of atom world lines.

more than a seething mass of never-ending world lines, continuous threads in the fabric of spacetime.

What does it mean that your body has nothing in common with the body you had a few years ago? If you are something other than the collection of atoms making up your body, what are you? You are not so much your atoms as you are the *pattern* in which your atoms are arranged. For example, some of the atomic patterns in your brain code memories. Figure 5.11 is a reminder that people are persistent spacetime tangles. To simplify the diagram, a person is represented by a set of four atom threads that have come close together. (An "atom thread" is the spacetime trail of an individual atom.) Note that an atom can leave one person's array and become part of another person. Very likely you have an atom of Jesus of Nazareth coursing through your body.

What I find most striking about Figure 5.11 is that the boxes enclosing individual people are completely imaginary. Mathematician Rudy Rucker has noted, "The simple processes of eating and breathing weave all of us together into a vast four-dimensional array. No matter how isolated you may sometimes feel, no matter how lonely, you are never really cut off from the whole."[1]

Light Cones and Omnipresence

Just as world-line diagrams are used to visualize the relationship between motions through spacetime, light-cone diagrams are also helpful in under-

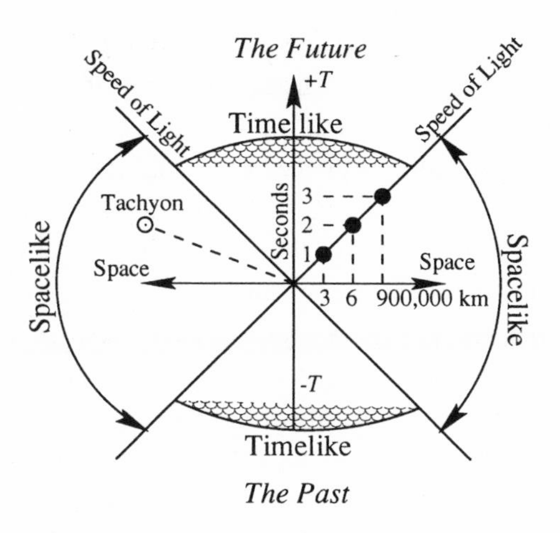

Figure 5.12 Light cone diagram.

standing the possibility of time travel. In Figure 5.12, we see a four-dimensional spacetime represented by two lines, one for space and one for time, much like in Figure 5.3. The vertical direction represents time. The horizontal direction represents space. Even though we normally think of space as having three dimensions, here we represent it by a single direction (horizontal) for clarity. Passage of time into the future is represented by upward movement, above the horizontal line. Everything that happened in the past is below the horizontal line. The point at which the time and space axes meet represents your "now." Consider it the starting point for any event.

We can use this figure to understand how a photon, a particle of light, would behave in spacetime. Light travels at about 300,000 kilometers per second. At this speed, one could circle the Earth about seven times in a second! If a photon starts at the "now" point, in one second it will be at the lowest dot on the diagonal line. In another second, it is at the second dot, and so forth. By connecting the dots, we create the world line of the photon that travels at the speed of light. Because the photon can travel in both spatial directions on this diagram, we draw a second diagonal line. These two diagonal lines encompass a region in which all sublight particles travel. Anything traveling in this region exhibits "timelike" behavior, and normal objects in our universe are confined to this area. In fact, most of the events in our universe cluster like crawling ants along the vertical time axis. In spite of these apparent limitations to motion, physicists do consider the existence of hypothetical particles, *tachyons*, which can move at speeds faster than light. Consider a tachyon that starts traveling just above light speed and then moves even faster. One such tachyon is represented at the left of the diagram. As the tachyon approaches infinite speed, it covers more space per unit time, and hence its position on the diagram grows closer to the space axis. Finally it moves along the horizontal space axis. It traverses space in no time at all. Particle behaviors that cluster around the space axis are called "spacelike." In a mystical way, we might refer to these infinite-speed particles as "omnipresent" because they simultaneously exist at different spatial locations.

Light-cone diagrams, such as the ones in this chapter, are useful but usually considered in the framework of special relativity, which is too primitive to settle the question of whether or not time travel to the past is possible. To answer the larger question, we must turn to general theory of relativity, which deals with ways in which spacetime can be twisted (for example, by gravity). We will see in future chapters that the full power of general relativity seems to permit looping world lines. There is considerable debate in the scientific community on whether the closed loops, or *closed timelike curves* (CTCs), exist. For example, curved space may tip light cones to create temporal loop-backs (travel into the past) without superluminal motion. Interestingly, the presence of any mass will tip a light cone, but the effect is small in our daily lives. However, an extremely massive body will cause a pronounced tipping of the light cone toward the body. If the body rotates, the cone will tip further in the direction of rotation. These effects as they apply to time travel will be discussed in Chapter 17 on the subject of Tipler cylinders.

Paradox and Parallel Universes

> You cannot underestimate the effect upon us of the creation of the physical universe. Think for one moment, if you can, what Time means, and how miserable you might be without it. No, that's not right. What I mean is, without Time you could not be conscious of yourself, either in terms of failure or achievement, or in terms of any motion backwards or forwards, or any effect.
>
> —Anne Rice, *Memnoch the Devil*

> When Matter was created, so was Time. All angels began to exist not only in heavenly perfection with God but to witness and be drawn into Time.
>
> —Anne Rice, *Memnoch the Devil*

If time travel to the past becomes possible someday, wouldn't there be crowds of time tourists at the crucifixion of Jesus and other important or popular events? If there were no future restrictions on time travelers, wouldn't the number in attendance at famous events equal the number of future time travelers for all time? Because popular historical events are not so crowded, some scientists believe this is evidence that time travel is impossible. Perhaps one or two time travelers could blend in with local crowds at historical events, but if many time travelers had been present, surely the historical record would document this. On the other hand, this doesn't mean that time machines will never be invented. For example, if time travel is achieved within general relativity by constructing a region of spacetime containing closed timelike curves, then time travelers could use these regions of spacetime for time travel into the past, but they would be restricted from going back to a time before the space warp existed. If the crucifixion of Jesus occurred before the existence of this kind of time machine, then this event could not be visited. However, once the time machine is invented, all events might be equally accessible after this point in time.

As discussed by scientists and science-fiction writers, various paradoxes could arise with time travel to the past. For example, if you accidentally killed your mother before you were born, how could you be alive to kill your mother? Or suppose that you traveled back in time and delayed your dad from having sex with your mother by just a few seconds. Then a different sperm would have penetrated her ovum, and you would not have been born. (Some believe these kinds of paradoxes won't arise because the very fact that you are alive at this moment means you will never go back and kill your mother before you were born!) On the other hand, paradoxes might be avoided if you could travel to the past and observe it without affecting anything. However, this could be quite difficult, because even the most tiny alterations in the past may become amplified through time. As some science-fiction writers have suggested, even the accidental killing of an insect could cause a ripple of ever-increasing consequences through time.

One way of avoiding paradoxes that could arise from backward time travel is the notion that the universe splits as a result of time travel. The new universe diverges from the old one, which follows the history the time traveler knew before going back in time. In other words, the time traveler creates an alternate, parallel universe by traveling back in time. Killing your mother before your birth means you would never have been born in this newly created universe; however, it does not change the fact that you were born in the universe from which you originated. As an example in science fiction, Mona Clee's recent novel *Branch Point* also resolves the time travel paradox by assuming the creation of a new universe as a result of time travel. In the book, the protagonists attempt to prevent a nuclear holocaust that occurred in their world on October 31, 1962, when John F. Kennedy invaded Cuba during the missile crisis. The crisis had escalated, nuclear bombs were used, and a nuclear winter was produced that destroyed most of humanity. The protagonists go back in time to try to create another universe where humanity flourishes.

Another, older example of paradox avoidance was published in David Daniel's 1934 story "Branches of Time," which similarly asserted that people could travel to any point in the future of their universe, but when they enter the past, the universe splits into two parallel worlds, each with its own time track. Along one track the world continues as if no looping has occurred. Along the other track, a universe is spawned and its history diverges from the original. From the standpoint of modern physics, the newly spawned universe is not truly new because a being living out of spacetime in a fifth dimension could observe the time traveler's world line switching from one spacetime continuum to another on a graph showing the universe branching like a tree in a "metauniverse."

To better understand "universe branching," consider an example. Suppose you go back to the time of Chopin in Universe 1 and cut off his thumb. The world forks. You are now in Universe 2. If you like, you can return to the present of Universe 2, a universe where the crippled Chopin had never become a great musician. In what way would this world be different from the original one? Would you find a duplicate of yourself there? Maybe, maybe not. As I just mentioned, several stories assume that the slightest alteration of the past would introduce new causal chains that would gradually diverge and finally produce big historical changes. It would be like giving a tiny push to a snowball causing it to roll down a mountain eventually producing an avalanche with a large effect. If Chopin never played piano, the lives of all of the people who attended his concerts would be subtly affected. Some children wouldn't be born. Certain people would or wouldn't be murdered. Chaos theory states that slight changes can produce amplified effects through time.

What would happen to you, personally, if you went back and chopped off Chopin's thumb? For one thing you wouldn't be annihilated for changing history; you would still exist because you are now an alien from Universe 1 living in Universe 2.

Let's return to the idea of small changes in history producing amplified effects through time. Imagine what would happen if Cleopatra had an ugly but benign skin growth on the tip of her nose. The entire cascade of historical events would be different. A mutation of a single skin cell caused by the random exposure to sunlight will change the universe. This entire line of thinking reminds me of a quote from writer Jane Roberts:

> You are so part of the world that your slightest action contributes to its reality. Your breath changes the atmosphere. Your encounters with others alter the fabrics of their lives, and the lives of those who come in contact with them.

In her novel *Memnoch the Devil*, Anne Rice has a similar view when she describes heaven:

> The tribe spread out to intersperse amongst countless families, and families joined to form nations, and the entire congregation was in fact a palpable and visible and interconnected configuration! Everyone impinged upon everyone else. Everyone drew, in his or her separateness, upon the separateness of everyone else!

In scenarios involving parallel universes, it is easy to create duplicates of yourself. For example, you can go back a year in Universe 1, live there for a year in Universe 2 with your replica in Universe 2, then again go back a year to visit two replicas of yourself in Universe 3. (One of my favorite tales of time travel replication is David Gerrold's *The Man Who Folded Himself.*) By repeating such loops you can create as many replicas as you like. The universe gets complicated, but there are no logical contradictions. Although this multiple-universe concept may seem far-fetched, serious physicists have considered such a possibility. In fact, Hugh Everett III's doctoral thesis "Relative State Formulation of Quantum Mechanics" (reprinted in *Reviews of Modern Physics*) outlines a controversial theory in which the universe at every instant branches into countless parallel worlds. However, human consciousness works in such a way that it is aware of only one universe at a time. This is called the "many worlds" interpretation of quantum mechanics. The theory holds that whenever the universe ("world") is confronted by a choice of paths at the quantum level, it actually follows both possibilities, splitting into two universes. These universes are often described as "parallel worlds" although, mathematically speaking, they are orthogonal or at right angles to each other. In the many-worlds theory, there are an infinite number of universes, and if true, then all kinds of strange worlds exist. In fact, some believe that somewhere virtually everything must be true. There is a universe where fairytales are true, a real Dorothy lives in Kansas dreaming about the Wizard of Oz, a real Adam and Eve live in a Garden of Eden, and alien abduction really does occur all the time. The theory also implies the existence of infinite universes so strange we could not describe them. My favorite tales of parallel worlds are those of Robert Heinlein. For example, in his science-fiction novel *The Number of the Beast* there is a parallel world that appears identical to ours in every respect except that the letter "J" does not appear in the English language. Luckily, the protagonists in the book have built a device that lets them perform controlled explorations of parallel worlds from the safety of their high-tech car. In contrast, the protagonist in Heinlein's novel *Job* shifts through parallel worlds without control. Unfortunately, just as he makes some money in one America, he shifts to a slightly different America where his money is no longer valid currency, which makes his life miserable.

Just as in the science-fiction stories involving parallel universes, the many-worlds theory suggests that a being existing outside of spacetime would see all conceivable forks, all possible four-dimensional spacetimes, as always having existed. How could a being deal with such knowledge and not become insane? A god would see all Earths: those where no inhabitants believe in God, those where all inhabitants believe in God, and everything in between. According to the many-worlds theory, there would be universes where Jesus was son of God, universes where Jesus was the son of the devil, and universes where Jesus did not exist.

Much of Everett's many-worlds interpretation is concerned with events on the submicroscopic level. For example, the theory predicts that every time an electron either moves or fails to move to a new energy level, a new universe is created. Currently it is not clear the degree to which quantum (submicroscopic) theories impact on reality at the macroscopic, human level. Quantum theory even clashes with relativity theory, which forbids faster-than-light (FTL) transfer of infor-

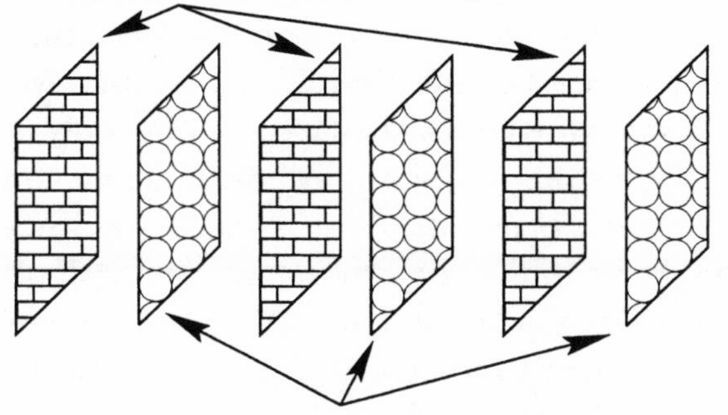

Figure 5.13 A bubbling river, located between time gaps in your universe, flows through the room in which you are watching television or reading this book.

mation. For example, quantum theory introduces an element of uncertainty into our understanding of the universe, and it states that any two particles that have once been in contact continue to influence each other no matter how far apart they move, until one of them interacts or is observed. In a strange way, this suggests that the entire universe is multiply connected by FTL signals. Physicists call this type of interaction "cosmic glue." The holy grail of physics is the reconciling of quantum and relativity physics.

What exactly is quantum theory? First, it is a modern science of the very small. It accurately describes the behavior of elementary particles, atoms, molecules, atom-sized black holes, and probably the birth of the universe when the universe was smaller than a proton. For more than half a century, physicists have used quantum theory as a mathematical tool for describing the behavior of matter (electrons, protons, neutrons) and various fields (gravity, weak and strong nuclear forces, and electromagnetism). It is a practical theory that can be used to understand the behavior of devices ranging from lasers to computer chips. Quantum theory describes the world as a collection of possibilities until a measurement makes one of these possibilities real. Quantum particles seem to be able to influence one another via quantum connections—superluminal links persisting between any two particles once they have interacted. When these ultrafast connections were first proposed, physicists dismissed them as mere theoretical artifacts, existing only in mathematical formalisms, not in the real world. Albert Einstein considered the idea to be so crazy that it had to demonstrate there was something missing in quantum theory. In the late 1960s, however, Irish physicist John Stewart Bell proved that a quantum connection was more than an interesting mathematical theory. In particular, he showed that real superluminal links between quantum particles explain certain experimental results. Bell's theorem suggests that after two particles interact and move apart outside the range of interaction, the particles continue to influence each other instantly via a real connection that joins them together with undiminished strength no matter how far apart the particles travel. Alain Aspect and his colleagues confirmed that the property is in fact an actual property of the real world. However, the precise nature of this FTL quantum connection is still widely disputed.

Chronon Theory of Time

So far we've been treating time as a continuous stream, but some physicists subscribe to the "chronon theory of time." In this theory, time is not continuous but made up of tiny particles jammed together like pearls on a necklace. The shortest time interval is the time for a quantum event (such as an electron slipping from an outer to an inner shell of an atom) to take place. Theoretically, such a time interval does not have a definite duration, but has only an approximate, unmeasurable size. The smallest definite time interval is the chronon, or 10^{-24} seconds (one million million million millionth of a second). This is the time it takes light to cross the smallest interval of space known to exist. In this theory, even though time may be discontinuous, we still perceive it to be smooth, just as we perceive movies to be smooth even though they are composed of a sequence of rapidly placed discrete frames. If the chronon theory is valid, then between each fundamental time interval there could be imperceptible gaps in which the basic units of time belonging to other universes could fit. According to chronon theory there might be an infinite series of real, solid universes stuck into the probability gaps between the quantum events of our own. If you are watching television peacefully in bed, there could be a mighty, bubbling river pouring through the time slices of an alternate universe (Fig. 5.13).

Imagine the chaos that would arise if time machines were as common as automobiles, with tens of millions of them commercially available. Havoc would soon break loose, tearing at the fabric of our universe. Millions of people would go back in time to meddle with their own past and the past of others, rewriting history in the process. . . . It would be impossible to take a simple census to see how many people there were at any given time.

—Michio Kaku, *Hyperspace*

Man's search for meaning is a primary force in his life.

—Viktor E. Frankl, originator of logotherapy, or existentialist analysis

It is interesting to note the curious mental attitude of scientists working on "hopeless" subjects. Contrary to what one might at first expect, they are all buoyed up by irrepressible optimism. I believe there is a simple explanation of this. Anyone without such optimism simply leaves the field and takes up some other line of work. Only the optimists remain.

—Francis Crick

Three Important Rules for Time Travelers

6

✍ "Constantia, where are you?" You nervously run your fingers through your hair and knock harder on the door to her sleeping quarters.

A minute later, she opens the door. She has a cup of coffee in one hand and a blow-dryer in the other. Her cheekbones are high and pronounced, her mouth tenderly curved, her eyes skillfully shadowed. The only imperfection she has (though you think it only adds to her allure) is an X-shaped scar near her left eyebrow—a memento of a childhood accident, she once told you. She seemed self-conscious about it.

"Hi, what's up?" Constantia says adjusting her green-and-gold denim dress. Her shirt is covered with little grey alien heads. Her shirt's buttons are shaped like flying saucers.

You stare at the faces on her shirt for a few seconds. "Today, we're going to talk about distant metrics and Minkowskian space."

"Sure, come on in. Mr. Veil is already here."

Mr. Veil waves to you. "Hello Sir, I was just finishing giving a violin lesson to Constantia.

She's learning very quickly."

"Mr. Veil, you and Constantia—"

"Sir, nothing to worry about. Just a violin lesson."

You are silent for a few moments and then walk into her room. You notice some changes since last time. This is still the room where she relaxes and reads and studies music and holds her stringed instruments close to her body. But now there are comfortable small couches fitted with heaps of pillows. There are halogen lamps of black iron so delicate and easy to maneuver that they look like praying mantises posed on tables.

You sit down on Constantia's bed. "Both of you, sit next to me."

When they are ready, you start to talk. "Today, I want to tell you how each of us perceives spacetime in relationship to one another. I'm going to prove the following:

1. All events in the future of your here–now are in the future of any other nearby, relatively moving person. The same applies to the past. In other words, all events in the past of your here–now are in the past of any other nearby, relatively moving person.
2. Any event in elsewhere can appear simultaneous with your here–now and not simultaneous for my here–now.
3. The order in time of causally related events is the same for all of us. In other words, the relations of before and after are the same for all of us."

You pause and say, "I call these the three rules that all time travelers should learn and prove. If you want to call yourself a scholar of spacetime, you must first understand the relationships between events in the future, past, and elsewhere."

"Sir, you can prove all this mathematically?"

"You can bet your ugly face on that, Mr. Veil."

"Sir!"

"Just joking Mr. Veil. Your multisegmented body still takes a little getting used to."

You look into Constantia's large, green eyes, but she offers no opinion. Her left iris sparkles with tiny liquid crystals.

You bring out a pad and pencil. "Let's begin and prove the first rule. Assume that Constantia is stationary as she observes an event happening at time t and position x. Recall from our discussions of the Lorentz equations that

$$t' = \frac{t - vx}{\sqrt{1 - v^2}}$$

$$x' = \frac{x - vt}{\sqrt{1 - v^2}}$$

I've set $c = 1$ to simplify the formulas. (Keep that in mind for the rest of the discussion!) t' is the time I measure in my reference frame moving at speed v. This means that

$$x'^2 - t'^2 = \frac{(x - vt)^2 - (t - vx)^2}{1 - v^2}$$

After a little algebraic manipulation the right hand side of the equation simplifies to $x^2 - t^2$.

"For Constantia in her stationary frame, an event is in her future cone if $t > |x|$, that is, if $t^2 - x^2$. This means that $x^2 - t^2 < 0$ for all of her future events. Now, notice that our previous result $x'^2 - t'^2 = x^2 - t^2$ means that $x'^2 - t'^2$ is also less than zero if the event is in the future. This proves rule 1: All events in the future for the here–now observer are in the future of any other nearby, relatively moving observer. We can use the same logic to prove the second part of the statement for past events."

Constantia gives your hand a squeeze. "Who would have thought you could prove that sort of time-traveling rule using simple mathematics?"

As she removes your hand, you feel a wonderful tingling sensation spread from your fingers up the length of your arm. You smile. For a few seconds, everyone is quiet.

Mr. Veil taps on Constantia's bed to get your attention. "Sir, what about the second rule?"

You nod and quicken the speed of your presentation. "Consider event *B*, the assassination of President Kennedy, and event *A*, the untimely death of thirty-nine-year-old Chopin in 1849." You sketch on the pad (Fig. 6.1). "Suppose that these two events occur such that Constantia, a

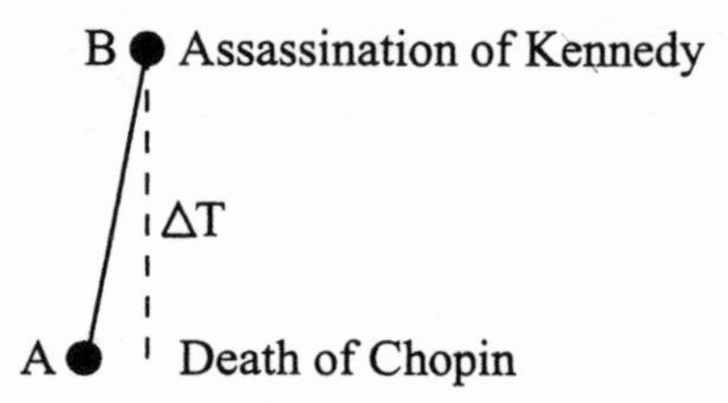

Figure 6.1

stationary observer, measures them to be $\Delta T = t_B - t_A$ apart in time. Using the Lorentz transformation, we have:

$$t_A' = \frac{t_A - vx_A}{\sqrt{1 - v^2}}$$

and

$$t_B' = \frac{(t_A + \Delta T) - vx_B}{\sqrt{1 - v^2}}$$

Combining these we get

$$\Delta T' = t_B' - t_A' = \frac{\Delta T + v(x_A - x_B)}{\sqrt{1 - v^2}}$$

Mr. Veil raises his hand. "Sir, simultaneity for the two events for the moving observer occurs at $\Delta T' = 0$."

"Correct, the moving observer therefore moves at the speed

$$v = \frac{\Delta T}{x_B - x_A'}$$

and this speed is less than the speed of light when the denominator $x_B - x_A > \Delta T$. For this to be true, event B is in the elsewhere of A. Thus, we have proven the second rule of time travel: Any event in elsewhere can appear to be simultaneous with the here–now for some observer and not simultaneous for other observers."

"Wonderful," Constantia says. She gives you a hug.

"Now for the final law. Let's consider event B to be in the causal future of event A. We have the condition then that $x_B - x_A < \Delta T$. Therefore,

$$\Delta T' = \frac{\Delta T - v(x_B - x_A)}{\sqrt{1 - v^2}} > \frac{\Delta T - v\Delta T}{\sqrt{1 - v^2}} = \Delta T \frac{1 - v}{\sqrt{1 - v^2}}$$

Recall that ΔT is the time between events. If ΔT is greater than 0, then $\Delta T'$ is also greater than 0 for $v < 1$. This proves rule three: The temporal ordering of causal related events is the same for all observers."

The three of you sit for a while, when suddenly Constantia jumps up. "Let's stretch our legs."

"Care for a walk?" you say.

Constantia adjusts her skirt and leads you and Mr. Veil out of the room and into a dimly lit atrium lined with mahogany walls. She goes up a flight of stairs, then past a long line of tiny offices—cubicles, really—where several museum employees tap endless information into desktop computers, talk on telephones, and shuffle musical scores. You wave to your colleague Aabel, a wasp-faced humanoid from the planet Eros whose fondness for gambling and alcohol has made him a pariah around the museum.

Then it is down another corridor, up another flight of stairs, and there you are, in Constantia's second lab. The tables are covered with violas and violins in various states of dissection.

You wrinkle your nose. The familiar, acrid odors of violin varnish, string cleaner, and sweat are unmistakable. Spiders have woven intricate webs over some violin strings, and in places the wood has splintered so much that it would fall away to powder at the touch.

"Constantia," you say. "You spend too much time in the museum. We've got to get out of here and spend some time in the real world. Tomorrow we'll have some more fun outside the museum."

The Science Behind the Science Fiction

> Science fiction, by including science, searches for new expressions of the basic grandeur and wonder of the world as it really is, as it can be understood through science and logic.
>
> —Eugene R. Stewart, *Skeptical Inquirer*, 1996

In this chapter, we've discussed the ways each of us perceives spacetime in relationship to one another. Many of the arguments rely on Einstein's relativity and the limiting speed of light. Einstein's theories concerning the nature of space and time are now commonly accepted facts of modern physics. Even after almost a century of experiments, not a single piece of evidence has ever contradicted the theory of relativity. The light-speed limit on objects seems to be built into the very structure of space and time.

The speed of light provides a convenient means for measuring both distance and time. Instead of expressing a star's distance from Earth in miles, we can measure it in "years," that is, the time it would take light to travel from the Earth to that star. The "light-year," therefore, is a measure of space expressed in units of time. The nearest star to Earth is Alpha Centauri, four light-years away. Our Milky Way galaxy is 100,000 light-years across. Obviously, these long interstellar distances are a major obstacle to exploration of space. In fact, the immenseness of space stimulated C. S. Lewis, author of the science-fiction series Perelandra, to suggest that this was God's quarantine mechanism for separating fallen worlds like Earth from other pristine planets. In Lewis's stories, space travel is a blasphemous attempt to circumvent God's will.

Light moves at the impressive speed of 186,000 miles an hour, traveling from the Moon to Earth in little more than a second. Saturn is an hour away by light beam. Because radio signals travel at light speed, signals from the Pioneer spacecraft near Saturn took at least an hour to reach Earth.

The light barrier may seem to be of no consequence to most of our daily activities, nor to our automobiles and other modern appliances. However, one machine is coming up against the Einstein limit—the computer. As computing speeds become faster, the overall operation speed of new computers will be limited not by processing rates of individual chips but by the light-speed-limited time it takes to transfer data between and through chips. For more than thirty years engineers have been building smaller and smaller silicon chips through which electrons pass. In 1990, however, researchers at

Figure 6.2 AT&T researchers shown with optical digital processors. As computing speeds become faster, the overall operation speed of new computers will be limited not by processing rates of individual chips but by the light-speed-limited time it takes to transfer data between and through chips. (Photograph reprinted with permission of AT&T. ©1990 AT&T.)

AT&T Bell Laboratories in Holmdel, New Jersey, developed the first optical digital processor that uses not electrons but laser beams to carry information. The Bell Labs machine may lead to computers in which the only limit to computing power will be the speed of light. In 1987, Bell Labs' physicist David Miller replaced silicon transistors on a microchip with infinitesimal mirrors. In 1990, Michael Prise was able to combine 128 of these "optical transistors" onto a single processor (Fig. 6.2). Prise used four arrays of 32 optical transistors—each array small enough to fit into the typed letter O.

The light-speed limit of computers has been overcome in several science-fiction scenarios. For example, science writer Richard Grigonis has suggested a fanciful "sixth-generation computer" whose chips are wired together with FTL links. Because of the time-travel implications of FTL signal transfer, a sixth-generation "Grigonis machine" could print out answers to questions before the questions are asked. Imagine how this machine would accelerate the pace of science!

"We're like millions of strands of spaghetti in the same pot. No time traveler can ever meet another time traveler in the past or the future. Each of us must travel up or down his own strand alone."

"But we're meeting each other now."

"We're no longer time travelers, Henry. We've become the spaghetti sauce."

—Alfred Bester, "The Men Who Murdered Mohammed," 1967

Losing control of the search is like casting aside maps, charts, and native guides in favor of thrashing about randomly. Columbus may not have known what he'd found, but he knew how to get there and back again.

—Eugene R. Stewart, *Skeptical Inquirer*, 1996

Existence is bound up with time. To contemplate the latter is to add to the enigma of the first. Is time the cosmic matrix of existence, or is it generated by it? Does time stand still, or does it flow? If it is in a state of flux, what is its speed? And since we measure speed by the ratio of traveling distance to the periodic motion of the clock, how are we supposed to measure the velocity of time? By time itself?

—Paul Hartal

Your Space or Mine?

7

✍ Constantia is looking at a map of Manhattan. Her shirt is covered with life-like drawings of Sergei Kovalev, a nineteenth-century Russian leader. Kovalev had criticized Russia's census-taking by saying it was a plot of the devil to increase his list of damned souls. Dozens of Kovalev's followers had themselves buried alive so they wouldn't be damned.

"Ready for our walk in the city?" Constantia says.

"Certainly."

Except for the wind on the museum windows, it is quiet. The thought of the cool wind sweeping toward you across Fifth Avenue is invigorating, and you feel stimulated by it. Constantia, Mr. Veil, and you exit the museum and head north along Fifth Avenue. A street vendor is selling pretzels and hot dogs, and you buy snacks for the three of you.

These days New York City is a free city, a galactic meeting place where aliens from Alpha Centauri can rub elbows just as easily with human soldiers patrolling the Van Allen belt as they can with housewives and smugglers.

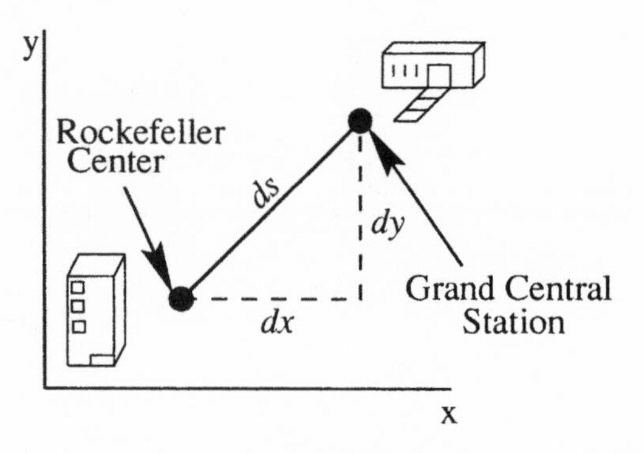

Figure 7.1 Spatial distances can be determined using the Pythagorean theorem for the lengths of a right triangle. The space distance metric is invariant with respect to the coordinate system because it does not depend on how you happen to select the *x* and *y* axes. You can rotate them and translate them, and the distance *ds* should be the same.

Constantia winds a long scarf around her neck. "How far is it to Rockefeller Center?"

"A few blocks." You pause. "Let's continue our lessons. Today I want to talk about distances in spacetime. I want to tell you about *spacetime intervals, Minkowskian space,* and *null intervals.*" Stooping down, you motion to a little alley next to a jewelry shop that has the words GOING OUT OF BUSINESS on the store front. You whip out a piece of chalk and draw two dots on the sidewalk (Fig. 7.1). "Here is Rockefeller Center, and here is Grand Central Station. In my drawing, both axes represent space. We can define a *distance metric* in terms of differential movements between Rockefeller Center and Grand Central Station. In the figure, if I make a differential movement along the coordinate axes of dx and dy, then the differential distance ds is given by

$$(ds)^2 = (dx)^2 + (dy)^2$$

I'm just using Pythagoras' theorem for the lengths of a right triangle. Before extending this distance metric to include time, I want to show you that the space distance metric is *invariant with respect to the coordinate system.*"

As you talk, a homeless man in a ragged coat comes by and listens. From across the street, you see a policeman staring at you with a questioning look.

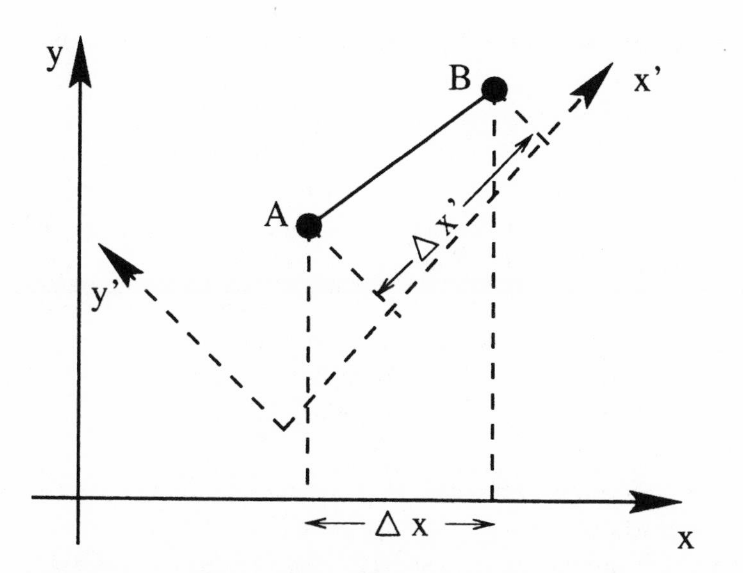

Figure 7.2 Rotated coordinate system. The Pythagorean distance function is invariant, the same for all coordinate systems.

You continue. "For example, this means that the distance on my map between Rockfeller Center and Grand Central Station does not depend on how I happen to select the x and y axes. I could rotate them and translate them, and the distance *ds* should be the same, although the (x,y) coordinates in the new system can change."

You squat down and draw another diagram showing a rotated spatial coordinate system with various distances in space (Fig. 7.2). The homeless man, stinking of perspiration and wine, also squats down beside you. You ignore him for the moment and point to the diagram. "Even though I've rotated and translated the (x,y) axes to make a new set of (x',y') axes, we find that $(dx)^2 + (dy)^2 = (dx')^2 + (dy')^2$. This means that even though the coordinates for A and B are different in the two systems, the distance between them is the same."

Constantia backs away from the homeless man and says, "This means that the distance function is *invariant,* the same for all coordinate systems."

"That's right," you say. "Now let's add time to the equation and see if distance is invariant in four-dimensional space time. If we consider all four dimensions, we can write *ds* as: $(ds)^2 = (dt)^2 + (dx)^2 + (dy)^2 + (dz)^2$." You pause and look at Mr. Veil. "Let's perform a rotation similar to the one we applied to the spatial axes, and think about the distance (ds') in the rotated system. I ask you, Mr. Veil, is it true that $(ds)^2 = (ds')^2 = (dt')^2 + (dx')^2 + (dy')^2 + (dz')^2$?"

Before Mr. Veil has a chance to consider your question, the homeless man grabs the chalk and says, "I hope you don't mind me interrupting, but I think it's easy to think about it if you consider just one space axis and one time axis. This means you're asking whether $(dt)^2 + (dx)^2 = (dt')^2 + (dx')^2$? Well, I can assure that this is not true." The man brings out a wine bottle, and you gasp. He continues. "If you use the Lorentz transformation, you can see that you can't extend the Pythagorean distance equation to two-dimensional flat spacetime."

The policeman comes over to watch. You grab back the chalk from the wino whose beard has been yellowed by nicotine and whose collar is rumpled and greasy. "You are right," you say grudgingly to him. "Perhaps it's because that the three *spatial* dimensions are orthogonal, that is, perpendicular to one another. But how can we make a time axis orthogonal? We can try to represent it in an imaginary direction, with $i = \sqrt{-1}$ to produce $(ds)^2 = (idt)^2 + (dx)^2 + (dy)^2 + (dz)^2 = -(dt)^2 + (dx)^2 + (dy)^2 + (dz)^2$. I realize that using imaginary time might seem to you as something out of *Star Trek*, but notice how this has changed the sign of $(dt)^2$."

The disheveled man says, "Damn important change—because your new metric is invariant."

You look at the man. "Who the hell are you?"

The man pays no attention to you. You again try to ignore him, perhaps because the man looks a little like you, except older. Gives you the creeps. Thank God you were fortunate to have a better life. "For convenience," you say, "I'm just going to use the negative of the metric. I'll explain why tomorrow. This give us $(ds)^2 = (dt)^2 - (dx)^2 - (dy)^2 - (dz)^2$." Now you gesture to the man. "If you use this simplified version where space is represented by a single axis x, this reduces to $(ds)^2 = (dt)^2 - (dx)^2$. This is the *distance metric* for spacetime."

You stretch your legs and continue your lecture. "We can use Lorentz's equations for x and t and show the invariance of these quantities. That is $(dt')^2 - (dx')^2 = (dt)^2 - (dx)^2$. The quantity on either side of the equals sign is called the *spacetime interval* between two events separated in flat spacetime by either dt, dx, dy, and dz or by dt', dx', dy', and dz'.

Constantia says, "You mean that $(ds)^2 = (ds')^2$."

"Yes!" you reply as a pigeon dropping lands on your spacetime diagram. "This means that if you are in the primed system and I am in the unprimed system we would see different individual space and time separations for two events, but we would see the same *interval!* The spacetime we're discussing here, with an imaginary coordinate for the time axis, is called *Minkowskian.* Because we're considering spacetime to be flat, where for every straight-line geodesic there are infinitely many parallel to it, we call it *Flat Minkowskian* spacetime. But in the real world, spacetime is not flat—gravity does strange things to it, and real-world spacetime does not follow the geometry of a flat piece of paper. Flat Minkowskian spacetime is a special case of what's called curved *Riemannian geometry.*"

Constantia shakes her head as if she has trouble grasping all the new facts.

You continue. "But since Riemannian geometry is even harder to understand, I find it easy to start with concepts of flat spacetime, to see what is possible in a simplified system."

The policeman comes closer and looks suspiciously at your drawings on the sidewalk. You notice that his nose has a squashed look, as if it were somehow contracted by a Lorentz transformation. "People," he says, "could you move it along? No loitering allowed."

Constantia smiles at the policeman. "Just one more minute, officer."

You clap your hands to get Mr. Veil and Constantia to pay attention. "I want to finish this lesson with a particularly interesting case. Recall that the spacetime interval ds is $(ds)^2 = (dt)^2 - (dx)^2$ where t is time and x is space. Now I want you to consider the special case of photon being emitted at one point in space and absorbed at another. What is the interval between the events?"

The police officer brings out handcuffs and moves toward you. "Please move along," he says.

Without even looking at the policeman, you begin to head north along Fifth Avenue as Constantia and Mr. Veil follow you. Fifth Avenue at this time of day was still a bustling thoroughfare with hovercrafts, motor scooters, delivery trucks belching diesel fumes, and Zetamorph bikes resembling pentagons with wheels. Every now and then you pass a teenager who has grown a third earlobe from which she or he can dangle additional earrings. These days, genetic engineering and reconstructive surgery are common.

The homeless bearded man follows Constantia, and you immediately place yourself between her and the man.

Without warning, the bearded man reaches over and squeezes your upper arm, checking your biceps. "Still in good shape," he says.

You stop walking. "What are you doing?" His hand remains on your biceps. It is making you uneasy. You don't like being touched by men you don't know. You back away from the man and turn your attention to Constantia.

You continue your lecture. "It turns out that the interval between the photon absorption and emission events is zero. In fact the interval is zero for any events connected by light. For Minkowskian spacetime we can take our definition of interval and take the derivative of each term with respect to time:

$$\left(\frac{ds}{dt}\right)^2 = 1 - \left(\frac{dx}{dt}\right)^2$$

This simplifies further. We know that $(dx/dt)^2 = 1$ because a photon travels at the speed of light, and this means that $(ds)^2 = 0$."

Mr. Veil scratches his bulbous head. "You mean that the world line of any photon has a zero interval?"

"Yes, it's called a null interval, and null interval world lines are always on the surface of light cones. Timelike intervals on the inside of light cones with $(dx/dt)^2 < 1$ have positive intervals $(ds)^2 > 0$. Outside the light cones are spacelike world lines with negative intervals (since $(dx/dt)^2 > 1$), which means that $(ds)^2 < 0$."

You are approaching Central Park. Constantia smiles. "I notice one big difference between distance measured in space and measured in spacetime. In space, the distance is always non-negative."

You nod. "Here's a weird result of spacetime null intervals. Consider three points A, B, and C at spacetime coordinates of (1,4), (3,2), and (1,0) respectively." You sketch on the pavement (Fig. 7.3). "The interval

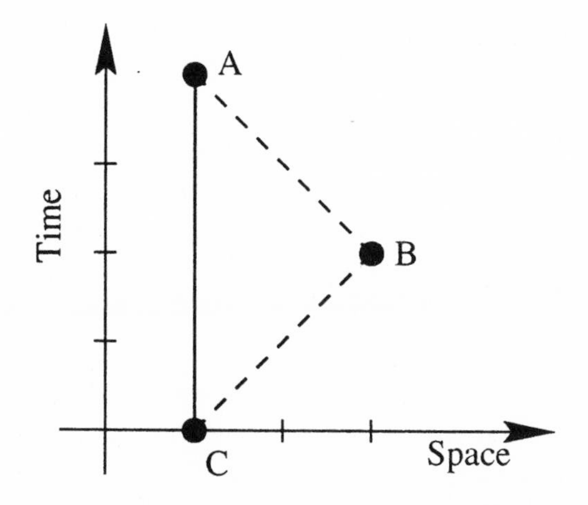

Figure 7.3 The weirdness of spacetime intervals. The interval between A and B is zero. The interval between B and C is zero. The interval between A and C is not zero.

between A and B is 0 and between B and C is 0, but the interval between A and C is not zero! The AB interval is a null interval for a particle traveling at the speed of light. It has a −45 degree slope. CB is also a null interval for a particle traveling at the speed of light."

"Let's take a break," Constantia says. "My head is swimming with all these new ideas."

"Good idea," you say, gazing into her sparkling eyes. For a second, you imagine photons traveling from her eyes to yours, joining them in virtual contact, separated by zero spacetime distance, the null interval. Your eyes are separated in spacetime by zero . . .

In a few minutes, you arrive at the lake in Central Park and see various humans and aliens frolicking in the sunshine. Orange water in the lake ebbs and flows. Glittering fairylike nodes of life dance on its translucent surface and float to its ancient rhythms. You look close. Within the lake, Ganymedean jellyfish, ctenophores, and all manner of Silurian sea creatures romp and prance, their diaphanous tentacles and evanescent egg sacs floating close behind in the cold liquid. Here and there are Gorgonian corals coated with algae.

Constantia's hair looks like little world lines blowing in the wind. "C'mon," she screams.

She has found a frisbee and is playing with Mr. Veil who has difficulty grasping the frisbee with his strange hands. The frisbee is one of those old-fashioned ones, made of plastic, not one of the latest gyroscopically controlled versions. She smiles and throws the frisbee to you, and as the interval between you and the frisbee decreases, you realize there is hope for you and Constantia.

You decide to impress her with your frisbee prowess. As it flies by, you turn the toy upside-down and launch it at a 37 degree angle to the ground and make it rotate at five revolutions per second. It seems to dance in the air for several seconds without moving and then shoots right into Constantia's hand.

"Ooh," she says. "Where did you learn to do that?"

A movement in the lake catches your eye, and soon something resembling a green and red möbius ribbon with blue eyes and a pompadour of greasy hair comes out of the depths and wriggles closer to Mr. Veil. Jellyfish creatures with purple bladders float to the surface of the amber liquid. Seconds later a vast magenta mass, perhaps their mother, joins the jellyfish. Its myriad looping arms curl and twist like a nest of boa constrictors.

Ah, life is good, you think to yourself. Central Park. The lake. The clear sky. There's nothing like a day at the park. ✍

The Science Behind the Science Fiction

> Our hopes, dreams, and wishes combine into knots of faith difficult to unravel with blunt, clumsy words.
>
> —Eugene R. Stewart, *Skeptical Inquirer,* 1996

When time and space are measured in the same units, the expression for the square of the spacetime interval between two events has a particularly simple form: $(\text{interval})^2 = (\text{time separation})^2 - (\text{space separation})^2 = t^2 - x^2$. (The units can be made the same; for example, the time measurement in seconds can be converted to meters by multiplying it by the "conversion factor" *c*, the speed of light.) This simple formula shows the unity of space and time. Einstein's teacher Hermann Minkowski (1864–1909) was so impressed by this formula that he wrote, "Henceforth space by itself, and time by itself, are doomed to fade away into mere shadows, and only a kind of union of the two will preserve an independent reality." Today we call this union "spacetime." Spacetime is the stage upon which all events occur, from atomic to galactic realms. The invariance of the spacetime interval—its independence of the state of motion of the observer—compels us to recognize that time cannot be separate from space. Space and time are part of a single entity, spacetime.

Neither past nor future could be changed—they could only be discovered.

—Poul Anderson, *Past Times*

You cannot escape one infinity, I told myself, by fleeing to another; you cannot escape the revelation of the identical by taking refuge in the illusion of the multiple.

—Umberto Eco, *Foucault's Pendulum*

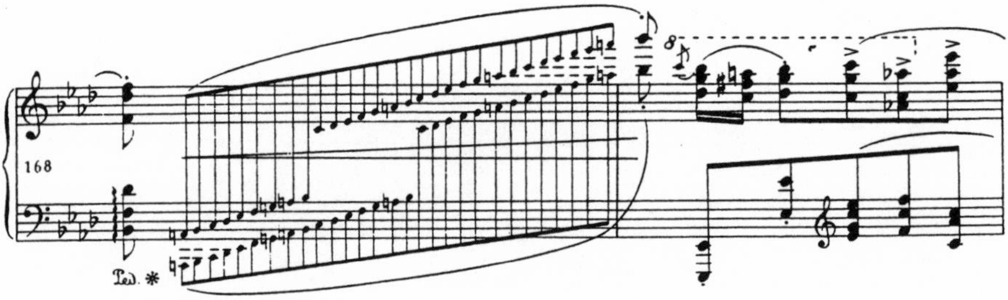

What then is time? If someone asks me, I know. If I wish to explain it to someone who asks, I know not.

—Aurelius Augustinus, Bishop of Hippo in North Africa

Alas, for every gram of knowledge we have about wormholes, black holes, time warps, parallel universes, and the light-speed barrier, we have six metric tons of unanswered questions.

—David Bauer, *Implosion*

How to Time Travel into the Future

8

✍ "Watch your step!" Mr. Veil yells as he looks at the floor.

"What?" you say.

"Golden-tailed geckos."

Without any warning, twin foot-sized creatures rear up their heads. You see their tails are covered with spiny protrusions. Their bodies are blue and brown, and speckled with little dots.

Mr. Veil jumps backward a few feet. "They can fire a sticky liquid three feet."

"Great," Constantia says sarcastically. She looks for more of the reptiles on the museum floor.

"Yes," Mr. Veil says, "I've seen them catch birds that way."

You nod. "I once saw a sage grouse fanning its tail feathers, popping the air from sacs in its throat to impress a hen, when suddenly the bird was covered with the sticky stuff. A bunch of geckos had it for dinner."

The pair of lizards scurry away, and you look out a window of the museum. It is raining, cool and light out of a hidden sky,

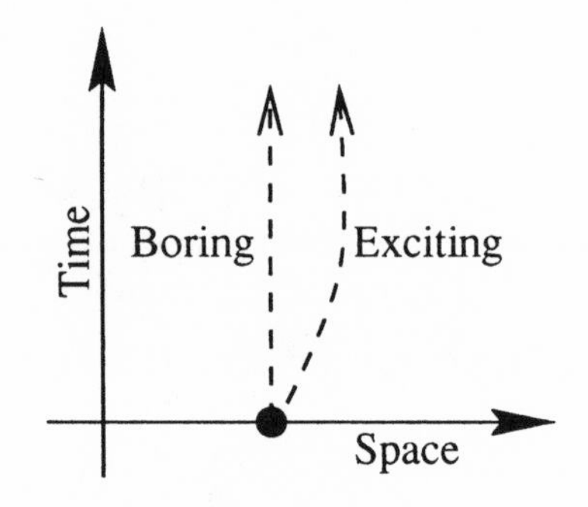

Figure 8.1 Two ways to travel into the future.

and the air smells fresh. You can just see the tall skyscrapers a mile away, under a floodlight glare.

You turn to Mr. Veil and Constantia. "Those twin lizards remind me of what we'll discuss today—curved world-lines, proper time, time travel into the future, and the twin paradox."

"Oh, that's easy," Mr. Veil says. "Just stand still, and you travel into the future". He draws a vertical world line on a spacetime diagram (Fig. 8.1).

"That's trivial. Here's a better way to travel *far* into the future. It's the most exciting result of relativity." You draw on his sketch.

"Can you explain?" says Constantia as she adjusts the collar of her shirt. You can't help staring at the center of Constantia's shirt, which has a photo of Michael Dorn, the actor who played Lieutenant Commander Worf on *Star Trek: The Next Generation.*

"Yesterday we talked about the flat spacetime metric, the 'distance' between two events separated in time (dt) and space (dx, dy, dz)." You draw an equation on the board: $(ds)^2 = (dt)^2 - (dx)^2 - (dy)^2 - (dz)^2$. "The fact that I'm using unprimed coordinates means that measurements of dt, dx, dy, and dz are made with respect to a stationary observer's reference frame. Now we're going to do something a little different. I want to make measurements of spacetime from the moving particle's point of view. In this new frame of reference, we have $dx' = dy' = dz' = 0$ because the particle is always at the origin of its own coordinate system.

"I also explained to you yesterday that the spacetime interval was the same, or invariant, for all observers. This means that

$$(ds')^2 = (ds)^2 = (dt')^2$$

for our coordinate system that moves with the particle. In other words, if you glue a clock onto the moving body, the spacetime interval between two events is the time lapse measured by the clock. This time interval is called proper time. Proper time belongs to—is the property of—the moving particle. The other day, one reason we used minus signs like $(ds)^2 = dt^2 - dx^2$ instead of $(ds)^2 = -dt^2 + dx^2$ was to avoid the result of imaginary proper time. For the moving reference frame, we didn't want the result $(ds)^2 = -dt^2$ since the space component is 0."

You look at Constantia and press a speaker button on the wall. Chopin's *Polonaise in C Minor* begins to fill the room. Constantia seems eager, happy, curious about you and your time travel lectures. No regrets about all the time she is spending with you.

You continue your lecture. "The rate of time-keeping (dt') of the accelerated clock depends on the rate of time-keeping of a stationary clock (dt) and the accelerated clock's instantaneous speed v:

$$dt' = \sqrt{1 - (v/c)^2}\, dt$$

It turns out that you can show that the total elapsed time between two events A and B, as measured by the proper time of the accelerated clock, is less than $t_B - t_A$, the elapsed time between A and B as measured by the stationary clock. This means that accelerated things age more slowly with respect to their unaccelerated partners."

You look out the window. The rain has stopped. "Let's continue our discussion in the fresh air. I'm in the mood for river sights." You carry a bag with props that will be useful later in your lessons. The three of you walk by a grand stairway leading up to the hall of wind instruments and then leave the museum. It's time for a long walk.

The streets to the re-engineered South Street Seaport wind down hill, following the kinks and bends of the river. The buildings are set well back from the streets, and it isn't until you get nearer to the water that you see there is a considerable number of people here. Thousands. The nearby avenues are themselves very busy with endless screams of noisy motor scooters and subsonic taxis often carrying whole families. It will take you another hour to get to the Seaport itself.

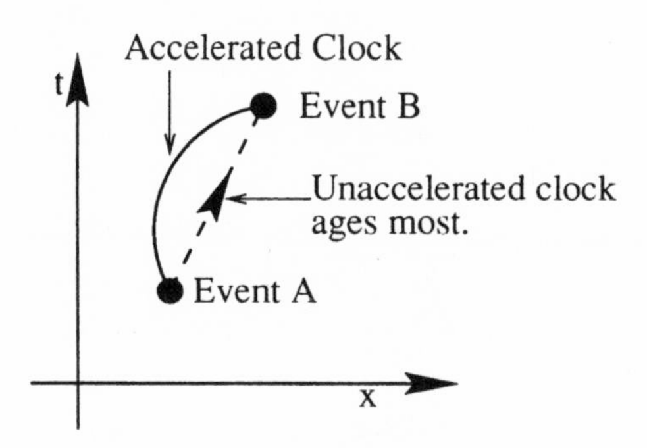

Figure 8.2 World lines of two clocks with different motions.

In the bright light, Mr. Veil appears almost like a small shriveled man, with a bald bulbous head above a thin gray face. Even on your informal outings, he wears a stiff-collared shirt and sober tie. One of the hardest things to take about him is the way his whip-tails nervously dance at the end of his multisegmented body.

"Sir," Mr. Veil says, "on spacetime diagrams, the world line of maximum proper time is the one that looks the shortest. It's a straight line."

"Right! Curved world lines look longer but they have smaller proper time. Any curved world line has a smaller proper time than the straight world line." You pause and sketch on the asphalt (Fig. 8.2).

Constantia says, "This result is strange. Quite different from the geometries of space and Euclidean geometry in which there is no longest path between two points."

You nod. "Minkowskian spacetime geometry is quite different from Euclidean geometry. This all leads to the famous time paradox sometimes called the twin paradox. For our example, let's use two brothers, President John F. Kennedy and Bobby Kennedy. Bobby stays on Earth but John is tired of being President so he climbs into a rocket and goes into space. At some point he turns around by firing his engines, and comes back home to Washington, D.C." You draw on the asphalt (Fig. 8.3).

"John and Bobby's world lines start at the White House, then split apart, and then come back together again at the end of John F. Kennedy's voyage. Bobby Kennedy's word line is straight. President Kennedy's world line is curved. Bobby's body will therefore measure a greater proper time than John's."

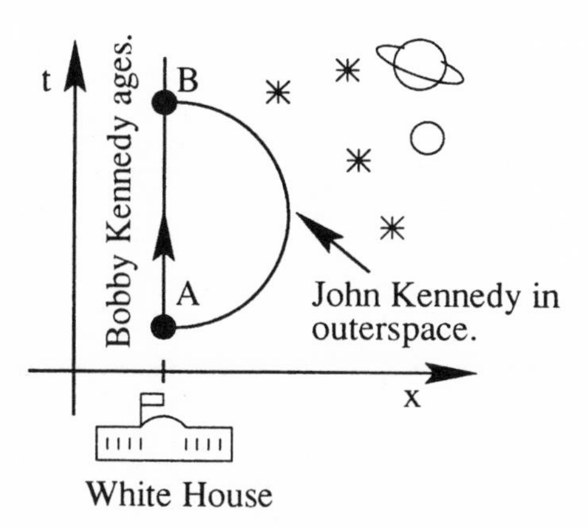

Figure 8.3 World lines of John (accelerated) and Bobby Kennedy (unaccelerated).

Constantia looks closer at your diagram. "You mean that John will be younger than his brother who stayed on Earth?"

"Precisely! When President Kennedy returns home, Bobby will tell him the date, and it will be further in the future than John's trip lasted (according to John). President Kennedy can only conclude that he has traveled into the future."

The three of you begin walking again. The city's fashionable shopping section has given way to a series of smaller shops. Many of them look a little seedy. You walk slowly, passing windows of secondhand clothing stores trying to pass themselves off as grunge boutiques. Some of the store signs read BUY PRODUCTS MADE ON EARTH. One of the stores is called Earthlings Unite and it sells a startling array of goods—handcuffs, skimpy nightgowns, and golden watches displayed on black velvet. A woman in the store, perhaps the manager, has a gaudy hairdo—half purple, half orange—and looks as if she might weigh around eighty pounds. She looks at the three of you and grins, revealing two large, white, canine teeth. They remind you of the teeth of some small but dangerous animal, a hyena, perhaps.

Constantia looks away from the woman and turns to you. "I understand now about future time travel by accelerating and returning to Earth, but why is it called a twin 'paradox'?"

"From Bobby's point of view, John is the one who travels away. However, one might argue from relativity that from John's point of view it is Bobby who is really moving away as John stands still."

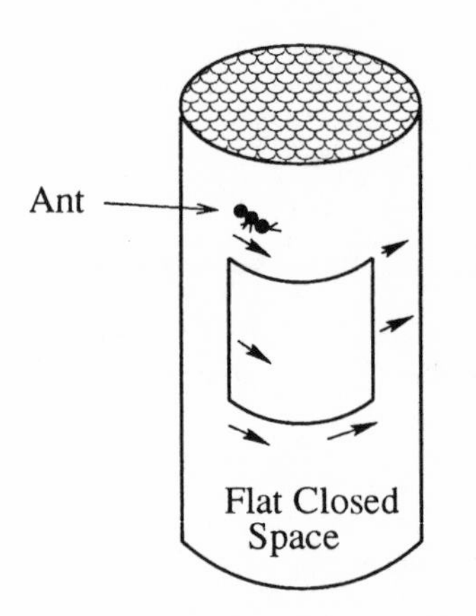

Figure 8.4 Ants on a flat, cylindrical surface. (The head of the ant is represented by an arrowhead.) If the ants walk along a closed path over the two-dimensional surface, their bodies do not rotate. Despite superficial appearances, this space is *not* considered curved.

Mr. Veil stretches his forelimbs. "So why does Bobby age faster than John?" he says.

"The answer is that the problem is not symmetrical, and it's not a paradox. It's John who feels the forces of acceleration due to the rocket's engine. Bobby doesn't. The best way to see the asymmetry is by looking at the spacetime plots. John's world line curves. Bobby's is straight."

In another half hour, you are walking with Constantia and Mr. Veil at the South Street Seaport. Constantia's suede boots resound hollowly on the dock—in sharp contrast to the peculiar tapping of Mr. Veil's feet.

Light from nearby ships glimmers off drenched wood. Constantia sits down on a bench and begins to pull off her boots to rest her feet. You sit down next to her, and edge closer, and she stays where she is. Her eyelashes flutter, long and smoky, against smooth, white cheeks.

You continue your discussion. "There are a few interesting complications of which you should be aware when thinking about world lines and acceleration. Curved world lines do always indicate acceleration in open flat spacetimes, but you can have *straight* world lines for accelerating bodies in *closed* flat spacetime."

Constantia's left arm brushes against yours. "What's an example of open flat space time and closed flat space time?"

"A flat piece of paper is a model for flat closed space time. If you roll the paper into a cylinder so that the ends meet, this defines a closed

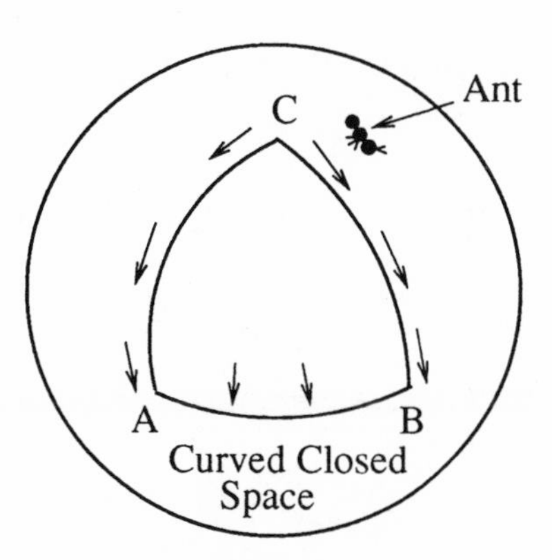

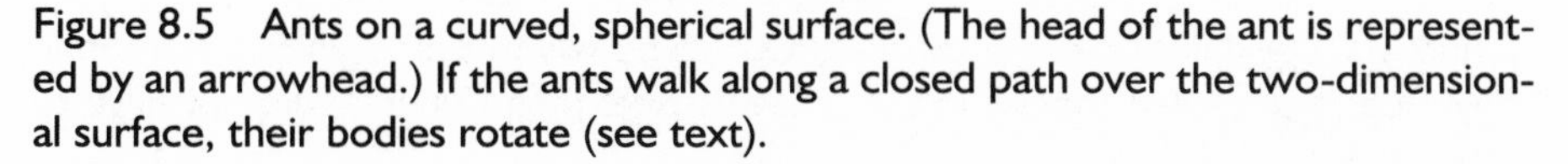
Figure 8.5 Ants on a curved, spherical surface. (The head of the ant is represented by an arrowhead.) If the ants walk along a closed path over the two-dimensional surface, their bodies rotate (see text).

space. The space is said to be flat because as an ant walks around a closed path while trying always to keep its body parallel to its previous position, the ant's body does not experience a rotation. The space is considered closed because an ant walking along a cylinder eventually can come back to itself by walking in a straight line—indicating the closure of its space" (see Fig. 8.4).

"Is the surface of a sphere a good model of closed, flat spacetime?"

From the small paper bag, you take out two balls and a can of Campbell's Chicken Noodle soup. You hand Constantia the balls and Mr. Veil the cylindrical can of soup.

"The sphere's surface is model of closed spacetime, but it's not flat. It's curved." You tap on one of the balls (Fig. 8.5). "Here's how you tell the difference. If the ant walks from *C* to *A* to *B* to *C*, all the while keeping itself parallel to its previous orientation with its head down, when it gets back to *C*, its head will point toward *B*, not towards *A*, as it originally did. The space is said to be curved because as the ant walks around a closed path while always trying to keep its body parallel to its previous position, the ant's body experiences a rotation. It has a non-zero rotation of its head. The curvature of the space is revealed by a process called *parallel transport.* On the other hand, the ant can walk around a cylindrical surface and return to its starting point with zero rotation of its body. So even though a cylinder's surface looks curved, it is not curved when considered as a model of space." You pause to take a deep breath and enjoy your surroundings. The dull brown color of the

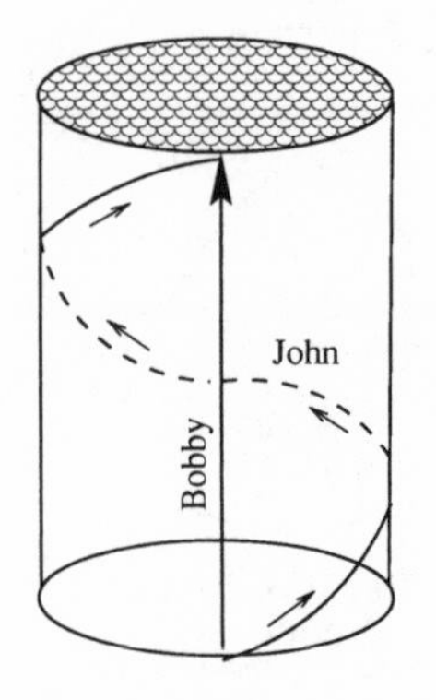

Figure 8.6 Unaccelerated twin paradox in a cylindrical spacetime. Here John Kennedy's world line is longer, so his proper time will be less than Bobby's when they meet again, even though in this closed space, John Kennedy did not have to accelerate.

wharf stands in contrast to the colorful beds of mutated flowers that line the walkways. Should you pick a flower and give it to Constantia?

Constantia is playing with your tennis balls, occasionally tossing one into the air and bouncing it against the other. "How's this all relate to time travel and acceleration?"

"As I was saying, curved world lines do always indicate acceleration in open flat spacetimes. For the journey of John F. Kennedy, his world line has to curve in order to accelerate and come back to meet his brother. However, let's model space using the surface of a cylinder." You sketch on the boardwalk (Fig. 8.6).

"Here we're using a two-dimensional surface as a simplified model of space. Both world lines are straight. If you don't believe that the helical line is straight, imagine unrolling the surface. When you do, the world line is straight. Here John Kennedy's world line is longer so his proper time will be less than Bobby's when they meet again, even though in this closed space, John Kennedy did not have to accelerate."

"Neat," Constantia says.

Mr. Veil is yawning. "It's late."

You look at your watch. "You're right. I think it's time to get back to the museum and go to bed. You've learned a lot today."

Much of the crowd near you at the Seaport has dispersed. A lone Chinese man in a magenta printed shirt and flip-flops walks by. He is walking his "Dat"—a transgenic cross between a dog and cat. Seems like most people these days want to experiment with Mother Nature now that genetic engineering is so easy.

Once back at the museum you and Constantia follow Veil up the stairs past his room, holding handkerchiefs against your mouths and noses. The smell of his place is much worse than you remember it. It is a heavy smell, a smell of putrefaction. Probably the smell of some of his leftovers or his latest cooking.

After saying goodnight to Mr. Veil, and then to Constantia, you return to your sleeping quarters and leave the door open, to listen to any sounds in the hallway. A museum curator is always alone. Maybe you should have married when you had the chance years ago. But the timing was not right. The person was not right.

You stare out the window. Darkness comes snaking across the museum like smoke. Outside there is a drizzle, and lamps throw broken yellow gleams off puddles. You are hungry, and find a can of raisins. It is so dark they remind you of little clumps of wrinkled world lines. Somehow the raisins don't taste right when invisible. You press a button, and Chopin pours into the room like water, water you have looked into, water you have held. ✍

The Science Behind the Science Fiction

> Reality might not behave always as we've come to expect. It's H. P. Lovecraft's crawling chaos, Nyarlathotep; it's the heart of a Stephen Hawking black hole; it's the core of Benoit Mandelbrot's fractals; in fact, it's Albert Einstein's assertion that, after all, physics as we know it might well be a local phenomenon.
>
> —Eugene R. Stewart, *Skeptical Inquirer*, 1996

Most physicists today believe that a high-speed rocket can function as a one-way time machine to the future. The situation you discussed involving John and Bobby Kennedy follows established relativistic theory. If John's rocket travels with a 1-*g* acceleration, he should be able to make significant penetration into the future. (A *g* is a unit of gravitational acceleration in which acceleration forces are measured. For example, a force of three *g* is equivalent to an acceleration three times that of a body falling near the Earth.) The journey should be comfortable because a *1-g* acceleration is equivalent to the Earth's gravity. To get a rough feel for the times involved during a 1-*g* trip, John can make a ten-year round-trip journey and penetrate twenty-five years into the future. If he stays in his ship for forty years, 59,223 years will have passed for Bobby on Earth! If John's body can withstand a sustained 2-*g* acceleration during his journey, he should be able to penetrate 906 million years into the future if he stays in his ship for forty years. (In the next chapter, I'll describe computer programs and formulas to computes these times.)

Time dilation is real, although small, even for slow speeds. When I drove to work today along the Taconic Parkway, in Westchester, New York, I experienced a time warp of about one one-thousandth of a nanosecond (assuming my average speed was 65 mph). This means that I experienced one one-thousandth of a nanosecond (a billionth of a second) less time passing than did my boss, who was sitting behind his office computer terminal, barely moving at all.

Want an example of a larger effect of motion on time? In the early 1970s, researchers placed atomic clocks on four commercial airlines. These cesium-beam atomic clocks were carried around the Earth, and they measured a 59-nanosecond (billionths of a second) time dilation when flying around the world in an eastward direction, and a 273-nanosecond time dilation when flying westward. Flying nonstop for a day had a time dilation effect of about a microsecond, or one millionth of a second. These effects were consistent with the velocity-related predictions of special relativity. (The east–west time difference is due to additional time dilation caused by the rotation of the Earth, as Einstein noted in his original paper that rotation of an object produces a time dilation too. Also, general relativity predicts that the higher-altitude clocks will record greater elapsed time due to the slightly reduced gravitational field, and these effects were taken into account when analyzing the results. Remember: Special relativity deals with velocity-dependent effects. General relativity deals with gravity-dependent effects.)

Here's another way-out trip to consider. Imagine that you climb into a rocket ship that accelerates (over a long period of time) close to the speed of light and travels to a star "50 light-years away." This means that light from the star would take fifty years to reach Earth. Does this imply that you should reach the star in just over fifty years if you moved close to the speed of light? Not at all. Fifty years would pass on *Earth* but not for you. In fact, for you in the ship, time runs more slowly than on Earth. If you traveled fast enough, it might take only a year for you to reach the star and return, even though according to your friends on Earth it would take 100 years for the round-trip. You would be returning to an Earth 100 years in your future.

Why are problems of this type called "twin paradoxes?" A casual analysis suggests that the high-speed rocket scenario is equivalent to a scenario where the rocket is considered stationary and the Earth is flying away from the rocket in the opposite direction. Given this way of looking at the problem, you'd expect the Earth clocks to run slower than the rocket clocks. Aren't the two scenarios identical? This is the crux of the so-called "twin paradox." However, the situation is not symmetrical. It is incorrect to think of the Earth as accelerating while the rocket stands still because the person in the rocket ship senses the acceleration and deceleration of the ship. Relativity does not make the two scenarios identical. Remember that the special theory of relativity applies to uniform motion, not to acceleration. An acceleration is not considered relative; it is absolute.

If you were to watch someone's clock on a spaceship with a powerful telescope while you sat on Earth, what would you see? What would they see? The person on the rocket should see the Earth clock running more slowly than expected by relativity. This extra slowing arises because the traveler does not see the Earth clock as it is at that instant, but as it was when the light left Earth some time before. The time taken for light to travel from Earth to the rocket will steadily increase as the rocket gets further from Earth. Thus the traveler will see events on Earth progressively more delayed, because the light needs to traverse an ever-lengthening gap between Earth and rocket. This nonrelativistic slowing of clocks and events is called the "Doppler effect," named after the Swedish physicist who first used it to describe experiments with sound waves. By adding the Doppler effect to the relativistic time-dilation effect, we get the combined slowdown factor that would contribute to the final visual appearance of the Earth clock. (This extra effect is usually discarded in discussions of the twin paradox.) Note that the Earth observer will also see the rocket clock slowed by the Doppler effect, because light from the rocket takes longer and longer to get back to Earth. The Earth observer will in addition see the rocket's clock slowed by the time-dilation effect.

Near light-speed travel of rockets would be very difficult for future civilizations for many reasons. For one thing, tremendous energy is required to achieve such speeds. The mass of the rocket increases with the speed, and a rogue asteroid would smash a ship to smithereens. Another obstacle arises because you have to travel quite close to the speed of light to get a substantial dilation effect. A relative velocity of one-third the velocity of light stretches time only by about 6 percent. Even if you could achieve 80 percent the speed of light in a small amount of time, this would imply an enormous acceleration, which could kill a human being.

When More is Less

In Euclidean geometry, a curved path has greater length, but in Lorentz geometry the curved world line is traversed in shorter proper time (i.e., wristwatch time or aging) than the more direct straight world line connecting two points. Because of this, different world lines between the same two events typically lead to different values of aging. Since a straight world line has the longest proper time, that aging is maximal along a straight line between two events. Even a quick kink in a world line decreases aging along that world line.

Let's work an example, so that you can experiment with aging on space-time maps of your own design. (In fact, this is a great classroom exercise

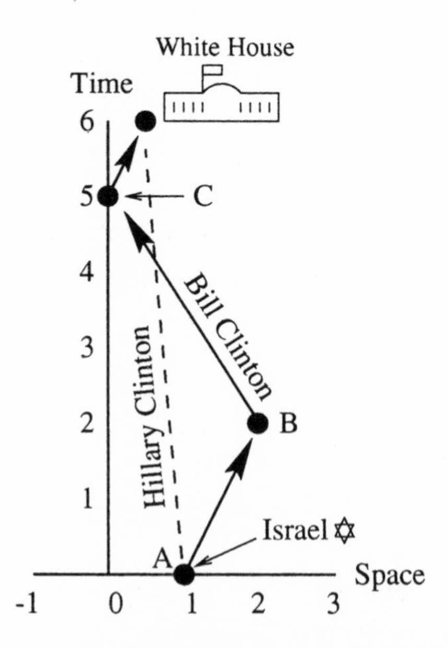

Figure 8.7 People age at different rates when traveling between the same two points but along different paths. Here Hillary Clinton ages more than Bill Clinton because she returns from Israel to the White House along a straight world line.

because students can design and plot various science-fiction scenarios and then visualize what time effects are involved.) Let us suppose that Bill and Hillary Clinton are on a trip to New Israel (a newly formed colony of Israelites in outer space) and they decide to return to the White House along different routes (Fig. 8.7). Bill Clinton moves along the solid straight world line segments from event *A* to events *B*, *C*, and finally to the White House. Hillary moves along a more direct route—the straight dotted word line from event *A* to the White House. Do Bill and Hillary age differently? To answer this question, we use the formula (proper time)^2 = (difference in time)^2 – (difference in position)^2. Let's look at Bill's path. From *A* to *B*, we have a time separation between events of two years, and a space separation of one light-year. Thus, the proper time lapse on a clock carried by Bill from *A* to *B* equals $\sqrt{2^2 - 1^2} = 1.73$. Similarly, the proper time lapse between *B* and *C* is 2.23 and between *C* and the White House is 0.866. Bill's aging along world line *A–B–C*–White House is the sum of the proper times along the individuals segments: 1.73 + 2.23 + 0.866 = 4.86 years. To compute Hillary's aging, we use $\sqrt{6^2 - 1^2} = 5.91$ years. Therefore she has aged about a year more than Bill because her direct (dotted) world line is longer and has more elapsed proper time than Bill's indirect world line. In Lorentz geometry, "less is more"!

Proofs of Time Distortion: Muons and Glittering Gold

Time distortions produced by objects going at near-light speeds have been observed by studying cosmic ray particles called muons. These particles are produced high in the atmosphere, and they plunge toward the Earth at speeds close to the velocity of light. Alas, their lives are short. Since they have a half-life of only about two microseconds, they die out rapidly as they travel to Earth. (I like to imagine muons as rain showers in which most drops evaporate before they have a chance to hit the ground.)

Scientists who observe muons find that the particles are not thinning out as rapidly as their normal half-lives would indicate. When the number of muons are counted on top of a mountain, and then in a valley, we find far too many muons surviving long enough to reach the lower altitude. The only explanation is that their half-lives are longer at high speeds—that is, their clocks run slow. In our frame of reference fixed to the Earth, moving-muon time becomes stretched (dilated), perhaps by a

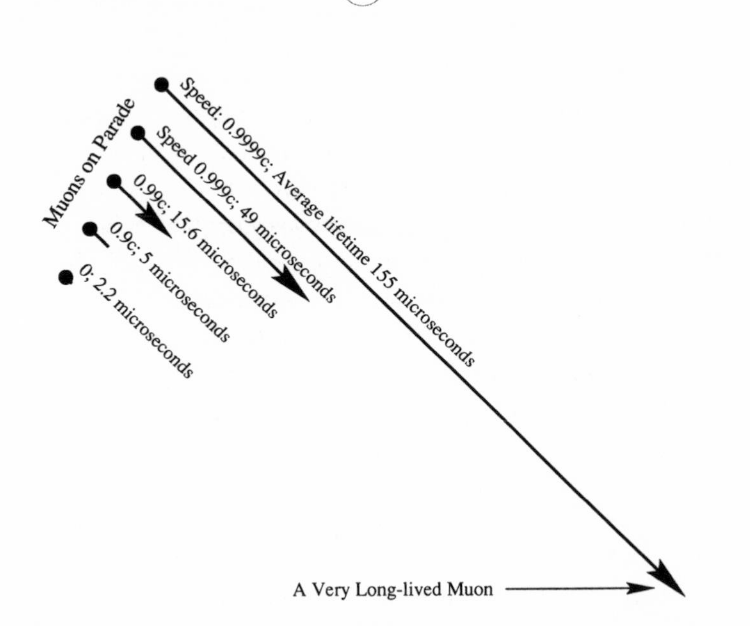

Figure 8.8 Muons on parade. The muon, symbolized by the Greek letter μ, is a particle that, on average, decays after about 2.2 microseconds into an electron, a neutrino (represented by the Greek letter ν, and an anti-neutrino $\bar{\nu}$). As the muon approaches light speed c, the stretching of time caused by relativity makes it take longer and longer to decay. This has been measured experimentally and is an important proof of relativity.

thousand times. Scientists observe muons by clicks on a Geiger counter, so in a strange way, the auditory signals are testimony that time travel exists. Physicists find this same kind of behavior exhibited by many different particles. As their speed increases, the lifetimes of particles also increase (Fig. 8.8).

The evidence for time distortion comes from many different experiments and is irrefutable. For example, in 1966, a group of physicists artificially produced muons using a particle accelerator and injected them into a ring-shaped vacuum tube where they circulated at 99.7 percent of the speed of light. This caused the muon's time to stretch by a factor of twelve relative to the lab, indicating the muons lived about twelve times as long as they would at rest. In 1978, researchers moved muons even closer to the speed of light and extended their lifetime by a factor of twenty-nine. Synchrotron electrons typically travel at 99.99999 percent the speed of light and their time dilation factor is as high as several thousand. These experiments leave no doubt that aging, clocks, and time are effected by motion.

One of my favorite pieces of evidence for time distortion involves the glitter of gold. For example, physicists take into account the time dilation of atomic electrons, including those near the nucleus, to get a complete understanding of the electrical and optical properties of materials such as gold. The distinctive, attractive glitter of gold results from the effects of relativity on the motions of the electrons inside the metal that are responsible for reflecting light. Paul Davies, author of *About Time,* notes, "It is no exaggeration to say that this precious metal is precious—and financially valuable—partly as a result of time dilation operating within the gold atoms."

"Time stretching" has also been confirmed with a device resembling a phonograph. In 1958, the German physicist Rudolf L. Mossbauer showed that certain atoms rigidly bound to crystals emit frequencies that are quite constant. Here's an experiment you can do with these emitting crystals. Place a crystal of vibrating atoms at the center of a turntable. These excited atoms produce gamma rays of a very stable frequency. On the edge of the turntable place another crystal of unexcited atoms that absorb the gamma rays. When the turntable is still, the absorbing crystal absorbs the rays. As the turntable increases speed, a detector will show decreased absorption because the atoms of the absorber begin to react to a different frequency due to its motion. The absorber will be operating on a different time scale from the atoms at the center of the turntable because they are revolving around it.

The search for faster-than-light communicators in the cracks of present-day physics has been compared to the nineteenth century search for perpetual-motion machines. In trying to understand clearly why perpetual motion machines invariably failed to work, physicists were led to the formulation of the first and second laws of thermodynamics which govern the amount and quality of energy available in any conceivable physical system. In a like manner, the study of why certain FTL schemes fail may also lead to certain general laws which on the surface seem to have nothing at all to do with the achievement of high-velocity communication.

—Nick Herbert, *Faster Than Light*

All things which can occur, do occur.

—Data, in *Star Trek: The Next Generation*

An insistence that there be room for ghosts, as well as for many other unexplained things, permeates humanity. We apparently want mystery, and we always have.

—Eugene R. Stewart, *Skeptical Inquirer,* 1996

Nothing travels faster than light, with the possible exception of bad news.

—Douglas Adams

Future Shock

9

✍ Constantia's eyes are dark and rather wide, as though she is lost in contemplation. Your eyes dart quickly around the Hall of Harps and return to her. You walk closer and notice that on her shirt are stenciled the words "Haight Ashbury, San Francisco." Some glass necklaces and wooden beads surround her neck like adoring snakes. From the corner of the room comes a thin trail of incense, and there is a strange musical sound coming from some nearby speakers. She is playing a CD of Bismilla Kahn and Ravi Shankar.

Mr. Veil is smoking on a pipe. His opal eyes are slightly dilated. "Sir, I think I found something that can travel faster than light."

You take a deep breath. "What's that?"

Mr. Veil withdraws a pair of shiny scissors and waves them in front of your eyes. "As I close the scissors, the notch where the two blades meet moves out toward the scissors' tip with a velocity that increases as the angle between the blades gets smaller. When the blades become parallel, the notch velocity approaches infinity. For very long scissors, the velocity of the intersection can exceed that of light during the last instant of closure."

"Mr. Veil, we know that the mass of an object increases as it approaches the speed of light. How could you close the scissors fast enough?" You shake your head. "Nice idea, but it could never work."

Mr. Veil points the scissors at you. "Wait. It's true that the scissors' blades have mass and must always move slower than light, but the intersection of the blades is massless. It's a mere geometrical concept, and is not subject to any relativistic mass increase."

Constantia begins to play on a konghou, a multistringed Chinese instrument of the harp family that first appeared in 206 B.C. Moisture from her hands is beginning to make the instrument glisten. She stops for a moment to examine the scissors. "Can the Einstein speed limit actually be exceeded by common cutlery?"

Mr. Veil opens his mouth wide, a peculiar grin revealing row upon row of bone-white teeth. "Imagine a pair of scissor blades extending to the moon, which is 1 1/2 light seconds away. If I close the blades in half a second, I can send a signal to the moon at three times the speed of light!"

"Mr. Veil, I'm sorry to say that your scissor-blade signaling scheme requires the scissors' material to be infinitely rigid, so that when you make a move on Earth, the entire blade moves at once. But that's impossible. There's no such material. If we did have such a material, you wouldn't even need a scissor to allow something to go faster than light."

"Sir, what do you mean?"

"For example, if you had a 'perfectly' rigid rod stretching from here to the moon, and you gave it a shove on Earth, the end of the rod on the moon would move at once—thereby breaking the speed of light. You could have used the rod as a faster than light communicator!"

Constantia nods as she continues to play on the konghou, her fingers moving like lemmings over the thirty-six strings. Suddenly her fingers slow down. "Can you make an infinitely rigid rod?" she says as she gently strokes the moist instrument.

You stare at her fingers as they make several glissandos and arpeggiated chords. "No. In reality, all rods are flexible, and deformations in such rods move at the speed of sound. In a real rod or scissors, a sudden movement on Earth would take a long time before it reached the moon. You can visualize this by thinking of a rubber rod to which you gave a shove."

Constantia stops her playing. "I think I have an example of something that can go faster than light," she says, her eyes bright as diamonds. "I've been working on it in my spare time." She puts away the konghou and brings out a board with dozens of little lights blinking in sequence, like the "moving" lights on the rim of a theater marquee. "I've programmed these lights to blink in various sequences. For example, I can program light 2 to blink the tiniest of fractions of a second after light 1, and so forth. This means that by proper programming, these lights can appear to move at any velocity, even velocities greater than light."

Mr. Veil screams in jubilant excitation. For a moment, his bulbous head appears to pulsate, and you fear it might explode.

You smile. "Excellent, Constantia."

She beams.

As you look at her and Mr. Veil, you realize how like children you all are at times. Give us praise and we'll keep trying for more praise. You know you would. You would work harder for praise than you would for money any day.

"But," you continue, "there's just one problem. Here's where your FTL machine fails. Nothing is really 'moving' at FTL speeds on the marquee, and certainly no message can be sent along your row of blinking lights. Einstein's theories don't prohibit superluminal motions or connections as such. What they do outlaw are FTL motions and connections that can act as a medium for causal influences."

You pause as you watch Mr. Veil's smile fade from his face. "If you build a light display in which light 1 triggered light 2 which triggered light 3, and so on, then we'd have a causal connection, and Einstein says that this type of light show can never move faster than light. In your device, you are driving all lights from a single source programmed to turn them on at specified times. Here the lights are not causally linked to one another. Einstein's prohibition doesn't apply, and any light pattern is possible, including patterns that ripple along at superluminal speeds! This same idea even applies to patterns on oscilloscopes and televisions. The speed at which a bright spot moves across the screen can, in principle, exceed the speed of light. In fact, scientists have discovered lots of wave phenomena that are superluminal, but, just as with the marquee lights, they arise in such a way as to be useless for FTL signalling."[1]

Mr. Veil stares at a komungo, a seventh-century zitherlike musical instrument from Korea, that sits on a nearby desk. You watch as he picks it up and begins to play it with his teeth.

"Mr. Veil!" you scream. "What are you doing?"

Mr. Veil moves the instrument away from his mouth. "Would you believe I was trying to imitate Jimi Hendrix?"

"No," Constantia says. "He is cleaning his teeth."

You shake your head. "Please put the komungo down. It's valuable. Now, pay attention. I want to tell you about a real example of FTL phenomena. Special relativity requires that all objects embedded in space obey the light-speed limit, but it places no restriction on the speed of expansion of space itself! Our universe appears to be expanding, and in fact large portions of faraway space are expanding faster than light. However, there seems to be no way to exploit this spacetime stretching for sending superluminal signals from one part of space to another."

You pause as you gently remove the komungo from Mr. Veil's forelimbs.

"Sorry, Sir, just wanted to make you laugh. Can we switch subjects?" His voice grows more serious. "I understand from yesterday's discussion that it's possible to travel into the future using a rocket ship. But how far you can travel into the future? I want to experiment on a computer."

"There's a convenient formula for doing just that. The relevant equation is $T = (c/a)\sinh(aT'/c)$ where a is the constant rocket acceleration and c the speed of light." You love pulling out impressive formulas like a magician pulling rabbits from a hat, but the formulas can all be derived using fairly simple math. "T is the time passage on Earth and T' is the time passage on the rocket. By 'constant acceleration,' I mean the acceleration as experienced by a person in the rocket ship. Observers on Earth would see the acceleration continually decreasing to zero as the rocket speed approaches the speed of light."

Mr. Veil grabs a notebook computer from a nearby desk and begins to type in a computer program (Code 2 in Appendix 2). His forelimbs are moving like a spider spinning a web.

You press a button on the wall and Chopin's *Polonaise in C Minor* fills the room. You continue. "Let me give an example using Frédéric Chopin. Imagine that he had a high-speed rocket available to him in the 1800s. To use the formula, let's assume that the trip is to be made in comfort, so that Chopin in the rocket experiences a 1-*g* acceleration. This would make him feel just like he would on Earth. Chopin travels for a time interval of T (as measured on Earth) and T' (as measured on the rocket). Then he turns off his rearward engine and turns on a forward-mounted engine so as to experience a constant deceleration. He does this for the same time interval and is brought to rest with respect to Earth. Then he returns to Earth using the same process. The total trip is $4T$ in Earth time or $4T'$ in rocket time." You pause. "Mr. Veil, are you done with your program? We should be able to show that $4T' < 4T$ and find out how far Chopin can penetrate into the future."

Mr. Veil hands you a computer printout of the results:

Years (rocket time)	Years (Earth time)
1	1.01
10	25.5
30	4,478
40	59,223
60	9,911,335

"Great Yggdrasill! If he stayed in the ship for sixty years, more than nine million years would pass on Earth!"

You nod. "Actual application of this method for future time travel would be difficult because such high rocket speeds would suffer a very high rate of collisions with atoms in outer space. For example, there is about one hydrogen atom in each cubic centimeter of space, and if many collide with your rocket there would be an intense lethal radiation in the form of gamma rays."

Constantia jumps up. "What the hell is that?"

Outside the window, you hear a commotion. The three of you peer out through the glass and see a crowd gathered around a lifelike statue of a man.

"It's another one of those FOPs," you say.

When you look at a person with FOP, or fibrodysplasia ossificans progressiva, you get the feeling that time is slowing down and finally stopping for them. They seem to move at different time scales as their tissues are converted to bone—transforming its sufferers into living statues. As the disease progresses, their spines, limbs, rib cages, and jaw bones fuse in place leading to complete immobilization in a bony prison. In some areas of New York City, they are left on the street as monuments, sad reminders that more funding is needed to find a cure for the disease.

Mr. Veil frowns. "It must be an epidemic. I saw a dozen FOPs while I walked down Forty-second Street the other day. One stood frozen in the atrium of Grand Central Station."

There is the sudden odor of creosote and lime. A tall cadaverous man with bushy eyebrows and bad teeth presses his face against the glass of the museum window, and you tell him to go away. He shambles backward and then trots away.

"Let's call it a day," you say as the sun finally goes down behind a mess of power lines and pylons.

"Wait a second," Constantia says. "You hear that?"

"Here what?"

"There. Scratching sound. You hear it?"

You stop breathing. "Yes, I think so."

Through the dusty window you see the lights of downtown New York City, a constellation burning through the thin fog, and just below, the peaked Chrysler tower rising majestically among humbler buildings, above the noise and the bustle of the city.

Suddenly, a large robotic mosquito thwacks against the window, attracted by the dim interior light. It clings to the glass, its long antennae swaying from side to side as if it were trying to eavesdrop on the conversation.

You wave at it to scare it away. "God, how I yearn for simpler times. Sometime I can hear Chopin playing in my dreams." You walk away from the window. "Let's get some rest." ✍

The Science Behind the Science Fiction

> The fairest thing we can experience is the mysterious. It is the fundamental emotion which stands at the cradle of true art and true science.
>
> —Albert Einstein, "The World as I See It"

For a derivation of the formula $T = (c/a)\sinh(aT'/c)$ see Nahin (1992). The term "sinh" refers to the hyperbolic sine, that is, $\sinh x = 0.5(e^x - e^{-x})$, and can be easily implemented in most computer program languages. You can experiment with the formula by changing the 1-g acceleration to higher values in order to see how this affects an astronaut's penetration into the future. For example, if you traveled for forty years at $2g$ (a=2), how far could you penetrate into the future? Do you think any humans would be alive on Earth to greet you upon your return?

Would you be interested in taking a high-speed trip so that you could see what the future holds? If so, how far into the future would you travel knowing that a return trip might be impossible for you? Certainly if you had a serious medical condition, time travel into the future would have appeal because future physicians might be able to cure you.

I don't think there's any question that a person could travel back in time while in a black hole. The question is whether he could ever emerge to brag about it.

—Princeton physicist Richard Gott

Eternity is very long, especially near the end.

—Woody Allen

Many animals can react to time. A rat can learn to press a lever that will, after a delay of some 25 seconds, reward it with a bit of food. But if the delay stretches beyond 30 seconds, the animal is stumped.

—Goudsmit and Claiborne, *Time*

Gravitational Time Dilation

10

✍ "Sir, there are government agents dressed in black in the museum."

"Mr. Veil, there you go again with your crazy ideas."

"They all have shaven heads. Black hats. CIA. I think they want your knowledge about time."

You are in Mr. Veil's living quarters, which are comprised of two adjacent rooms in the museum. The door through which you enter is windowless, with a fluorescent ceiling and ventilator grilles. A workbench supports mucilagenous scientific equipment, most of which you can identify. Upon his wall is a music library containing discs ranging from medieval chants to modern symphonies. His second room is the snappy, insubstantial style of the corporate world—lots of mahogany and leather and shades of beige and tan that could offend no one. The table and chairs were harmoniously Tibetan.

"CIA? That's nonsense." Your eye wanders over to a photo of an alien with a diamond body. "Who's that?"

"My old friend, Mr. Plex. Plays the harpsichord with lightning speed. Famous black hole researcher. He once told me about the time he went close to a black hole, and his wristwatch seemed to be in error when he returned to his ship."

"Mr. Veil, his watch was probably in perfect working order. Gravity causes time to slow down. Far away from the black hole, where spacetime is flat, clocks tick at their normal rates. Imagine you are on a ship watching your friend as he went closer to the black hole. You would see his movements slow down. As he moved into regions of increasing gravitational curvature, it would appear to people on the ship that his time slowed down."

"But he wouldn't notice it?"

"Correct, his heartbeat and thinking processes slowed down by exactly the same factor as his wristwatch."

Constantia, who had been quietly listening, suddenly says, "Does a clock on the ceiling of my room run faster than a clock on the floor? The clock on the floor experiences more gravity." On her shirt swim dozens of cartoon drawings of small, yellow squids.

You smile at her colorful shirt. "Yes, time flows more slowly near the floor."

Mr. Veil jumps back. "Great Yggdrasill! That means that the ants creeping on the ground are in a different time flow than humans."

"Mr. Veil, do you need a tranquilizer?"

"It's just—"

You pat Mr. Veil on his back. "The time difference is very small in your examples. But a black hole has a big effect. That's why I didn't send you into space and let you get close to a black hole. I didn't want you returning to the museum only to find me an old man."

"Uh, Sir, wouldn't you be concerned about my safety?"

You sit down on Mr. Veil's Tibetan rug so that you can look up at Constantia. She is perched on the edge of his bed, beige coat carelessly open now, and a streak of scarlet scarf showing, her face heavily made up in a way that makes her radiant and at the same time a little enchanting, as though she is not quite human. In fact today her hair looks a little smoother than human hair, the strands perhaps thicker, and certainly more iridescent.

Suddenly a gecko scoots by Constantia and she lets out a scream. The creature was about four inches long and beige.

"Don't worry about it," Mr. Veil says. "They keep the insects away. I like having them around."

You raise your hand to silence the Zetamorph. "Let's go find a nicer place where we can continue our discussion."

You lead them out of the museum and find a bar just a block away. As you walk, you pass by statue-people adorning the sidewalks like stalagmites. Some of these FOPs, these poor souls with fibrodysplasia ossificans progressiva, are immobile. Other black-clad FOPs cross the street with great difficulty and slowness, their faces hidden behind mirrored masks, and you shiver at the sight of these nighttime marchers, these rigid demons without faces.

Once in the bar, you find a small table cleaving to the wall. The place is not crowded. In one corner are several yellow-bearded men looking like Norse barbarians. In another corner, someone is playing a piano, very tenderly for a hotel bar, you think. And it is something from George Winston. What luck.

"Let's pretend you were to take a trip near a black hole. During one second of your time near the hole, millions of years could have flown past our museum. The equation that describes time flow is:

$$t_2 = t_1/\sqrt{1.0 - C_h/C^2}$$

Here t_2 is the elapsed time you experience when you hover close to the hole as compared with the elapsed time t_1 far from the hole. C_h is the circumference of the black hole's event horizon. (The event horizon defines a region of space surrounding the black hole. If you go too close and penetrate the event horizon, the gravity is so great that you can never escape.) C is the circumference around the hole at which you'd hover, slightly above the event horizon. If we experiment with the equation, we'd find that an observer near the black hole ages more slowly than he would in our cozy museum."

A bartender throws a hefty shadow over the small table, and murmurs something about drinks. You nod to him and send him away.

Mr. Veil shoves one of his forelimbs into a cavity of his abdomen and withdraws a notebook computer dripping with oleaginous body fluids. "Found a nice place to store this thing, Sir." He looks at the computer keyboard.

You let out a quick gasp. "Yes, you have."

Constantia jumps back. "Mr. Veil!"

He begins to type on the keyboard with several of his limbs moving together in rapid synchrony (Code 3 in Appendix 2).

"Sir, my program reports the following:

C/Ch	Time 1 (days)	Time 2 (days)
2	1	1.414213
1.5	1	1.73205
1.25	1	2.236067
1.125	1	2.999999
1.0625	1	4.123104
1.03125	1	5.744561
1.015625	1	8.062243
1.007812	1	11.35782
1.003906	1	16.0312
1.001953	1	22.6492
1.000977	1	32.01465
1.000488	1	45.26312
1.000122	1	90.50969
1.000031	1	181.0194
1.000008	1	362.0386
1.000002	1	724.0774
1.000001	1	1024

This means that if I hovered at 1.000001 times the event horizon circumference, one day for me would mean 1,024 days for you!"

The piano has moved on into something popular, from a Broadway musical, you think. It is sad and sweet, and one of the scantily clad women in the bar is rocking slowly to the music and mouthing the words with her rouged lips as she puffs on some strange weed. She is from the generations that had smoked so much that stopping now was out of the question. Still, she has the skin of a newborn baby.

You grin. "Mr. Veil, have you found the fountain of youth with your computer program?"

"I'm not sure I want to get that close to the black hole."

Your grin widens. "If you do go, just watch that cute little segmented butt of yours. If it were to slip beneath the event horizon while you hovered just millimeters above it—"

The Zetamorph takes a step back. "Sir, are you making fun of me?"

"Of course not." You turn to Constantia and say, "I think that's enough for today. We should start preparing a checklist of items to bring with us on our trip to see Chopin."

Rain starts to fall outside. The piano music grows more rapid and urgent: Bach's *Toccata* and *Fugue in D Minor* played with an African rhythm.

Constantia looks at you with wide eyes. "But you never told us how we will go back in time."

"Soon . . ."

A few minutes later you are alone in your sleeping quarters back at the museum. On the horizon, behind the skyscrapers, the moon is slowly rising. The big museum is filled with strange rustling sounds, perhaps mice, or the ghosts of ancient musicians. From time to time, you peek out your room. You turn on the *Polonaise-Fantaisie in A-flat Major, Op. 61,* abandoning yourself to the magic rolling of the arpeggios, accepting the revelation that your dreams of time travel are about to be realized.

You think you hear Constantia's steps running fast down the hallway away from you. All this talk of time travel makes you drunk as the formulas swim in your mind.

You climb out of bed, your head spinning, and stand for a while until the dizziness passes. But it is no good. Your limbs are tired. You sink back into bed, back into the dreams. Your eyes close so slowly they are like flowers closing.

The Science Behind the Science-Fiction

> I realized that if I understood too clearly what I was doing, where I was going, then I probably wasn't working on anything very interesting.
>
> —Peter Carruthers

Black Holes

The gravitational time dilation you discussed with Constantia and Mr. Veil is quite real. However, the effect is small in our daily lives. For example, over a lifetime, you could gain a microsecond or so over your high-rise apartment neighbors simply by living on the ground floor. Time on the sun's surface runs about two parts in a million slower than on Earth, because of the sun's much higher gravity. This effect was confirmed by American physicist Irwin Shapiro, who showed that radar signals and their reflections off planets are delayed when the sun is near the path of the signals.

An observer near a black hole can age much more slowly than one farther from the black hole. Also, the *proper time* of a clock on the surface of a collapsing star is different from the *apparent time* of the collapse, measured by a distant observer. This is because the surface is accelerating with respect to the distant observer. The contraction of the star below the Schwarzchild radius[1] happens in *finite* proper time, but in an *infinite* apparent time. You will never be able to see the formation of a black hole. You will see the collapse get slower and slower as light from the star becomes redshifted and fainter. (A redshift is a change in the light to longer wavelengths and lower energies as a result of the black hole's gravitation.)

What would happen if Mr. Veil approached too closely to a black hole? Like an ancient ant trapped in amber, Mr. Veil would appear permanently frozen at the event horizon, as his image gradually faded. In actuality, his body pierces the event horizon as it plunges into the singularity. (For additional information, see my book *Black Holes: A Traveler's Guide.*)

Gravitational Effects on Earth

As we've discussed, according to Einstein's general theory of relativity, the rate at which clocks keep time varies with the strength of gravity. What are some specific examples? If you were to ask Mr. Veil to climb to the top of the Empire State Building, where gravity is slightly weaker than on the bottom floor, a clock on his wrist would run slightly faster. The differences would be small, but since time (or frequency) can be measured more accurately than any other quantity, we can experimentally verify that for every ten meters' increase in elevation above the surface of Earth, gravity weakens by 0.0003 percent, and a clock would run faster by one second in 100 million years.

> On the upside, humans are so unadvanced when it comes to high-speed travel that the light-speed barrier doesn't even matter. Hell, just achieving one-100th the speed of light would be the greatest breakthrough in the millennium.
>
> —Dave Bauer, *Implosion,* 1996

> Although most physicists today place the probability of the existence of tachyons only slightly higher than the existence of unicorns, research into the properties of these hypothetical FTL particles has not been entirely fruitless.
>
> —Nick Herbert, *Faster Than Light*

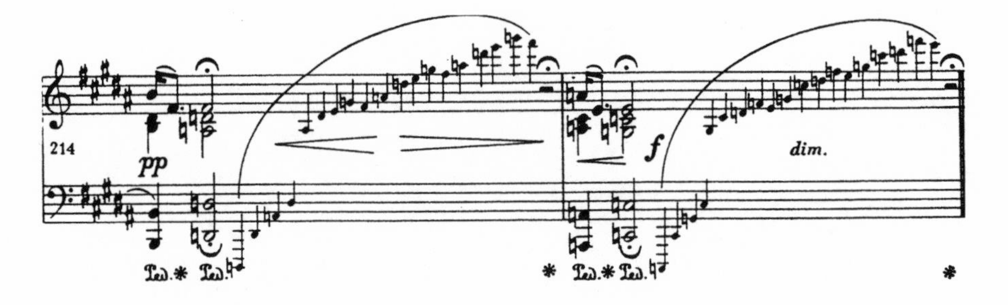

> If tachyons are one day discovered, the day before the momentous occasion a notice from the discoverers should appear in newspapers announcing "Tachyons have been discovered tomorrow."
>
> —Paul Nahin, *Time Machines*

> In relativity theory, in the subtle fusion of time and space known as Minkowskian space-time, the space dimensions seem to lord it over the time dimensions, and the whole structure exists as a frozen manifold outside of time.
>
> —Philip Davis and Reuben Hersh, *Descartes' Dream,* 1986

Tachyons, Cosmic Moment Lines, Transcendent Infinite Speeds

11

✍ "We can't take Mr. Veil back with us to see Chopin!" Constantia whispers as she wrings her hands together.

Today Constantia's shirt has a sketch of several Chupacabras, alien creatures once thought to exist in the Puerto Rican hills. The artist made their large eyes look comical.

"Why not?"

"It will cause too much commotion. There's no way you can disguise his Zetamorph face and body."

You are on the museum's grand stairway. It is one of those very opulent structures, divinely overdone, full of gold and crimson. From an open window, you see a hologram of St. Patrick's Cathedral across the street. Not the real thing. Too expensive to move it.

You and Constantia walk together onto the carpeted mezzanine and enter Constantia's room.

"Perhaps you are right," you say.

Constantia ties a length of shimmering crimson material into her abundant hair. It goes quite well with her sweater, a long skirt, and an elaborately embroidered waistcoat. "I feel bad for the guy," she says. "He's a pleasant fellow."

"Well, it would get a bit tight in the car with three people."

Mr. Veil enters the room and nods.

"It's okay, Sir. Maybe someday I'll travel myself into the future where they're more accepting of trans-species communion."

"Okay," you say with an deep exhalation of breath. "Now that we've cleared that up, today I'd like to discuss superluminal speeds, backwards time travel, tachyons, transcendent states of infinite speed, cosmic moment lines, and ultraluminal travel—"

Constantia laughs. "What a mouthful."

"First let's get some fresh air as we talk. I know a nice place near the Empire State Building."

The three of you leave the museum. On your way, you window-shop outside the various gadget stores. The Christmas lights and decorations have recently gone up, and you—along with hundreds of other equally mesmerized passersby—peer though the thick glass at scenes from Charles Dicken's *A Christmas Carol.* It is a robot version with robotic actors.

As you walk, you notice a scruffy, barefoot, dirty, gnomish man with blazing eyes and a greasy face. He seems to watch the three of you for a few minutes, but he soon runs away down the street.

You begin your lecture. "In the past few days, we've only been talking about speeds below the speed of light. In our equations, this meant that $v < c$ where v was the relative velocity of two reference frames."

"Can something really go faster than the speed of light?" Constantia says.

"Indulge me for a moment. In any case, the modern approach to special relativity emphasizes the invariance of the speed of light (like I explained to you before), and not that it is some limiting speed. I'll explain this more today. However, if v could be greater than c we can have a causality violation, because the mathematics indicates that an FTL particle travels into the past."

"Wouldn't that require infinite energy to penetrate the 'light barrier'?"

"Some scientists postulate the existence of FTL particles called *tachyons.* Einstein's relativity theory doesn't allow the acceleration of a massive particle to the speed of light, but it does allow a massless particle, a photon, to exist just at the speed of light. Scientists who think tachyons are possible suggest that they move faster than the speed of light at the instant they are created, so they avoid the problem of accelerating through the 'light barrier.'" You pause and bring out a pad of paper. "However, we still have some problems with this idea. The relativistic expressions for energy and momentum of a particle with rest mass m_0 moving with speed v are:

$$E = \frac{m_0 c^2}{\sqrt{1 - (v/c)^2}}$$

$$\text{momentum} = \frac{m_0 v}{\sqrt{1 - (v/c)^2}} \text{."}$$

Constantia puts her hands on her hips. "But the square roots in the denominators become imaginary if $v > c$. Doesn't energy and momentum have to be real-valued? After all, we can measure and observe the energy and momentum of particles."

"Very true. We can, however, make these quantities real-valued if we use $m_0 = i\mu$ for a tachyon. Here, μ is what's called a real-valued *meta-mass*, and i is an imaginary number equal to $\sqrt{-1}$. If we plug the value of this m_0 into the formulas, after some manipulation, we get:

$$E = \frac{\mu c^2}{\sqrt{(v/c)^2 - 1}}$$

$$\text{momentum} = \frac{\mu v}{\sqrt{(v/c)^2 - 1}}$$

Constantia stops walking. "What the hell is meta-mass? What could imaginary mass mean?"

"Those who believe in tachyons say that the rest mass of a superluminal particle would be unobservable because there is no subluminal frame in which the particle could be at rest. In other words, there's no frame of reference in which this mysterious imaginary mass could be measured. It's only observed changes in the real energy and momentum that characterize particle interactions."

Mr. Veil stares at the formulas. "The weird thing about the equations is they say that if tachyons lose energy, they speed up!"

You nod. "If there is a way for continuous energy loss, then tachyons accelerate without limit and enter what scientists call a *transcendent state of infinite speed*."

Mr. Veil nods to a few bearded dwarfs passing by, and then looks back at your pad of paper. "The equations seem to show that such infinitely speedy particles have zero energy but a finite momentum of uc."

You nod. "Yes."

You pause for a moment to view your surroundings. It is a thirty-minute walk from the museum to the Empire State Building. There's a lot of traffic on Fifth Avenue, so you cut through back streets and alleys, shouldering your way through the crowds on their way to work as you meander past scores of municipal and national government agency buildings, skyscrapers, streamlined office towers, and corporate headquarters from several worlds. Even the best city maps can't do justice to New York City's random pattern of streets and avenues near the museum—the place is a maze. For most people who don't live in New York, or have a hard time reading the street signs, the best thing to do is try to memorize landmarks and work from those.

Mr. Veil comes closer to you. "I'm still not sure how faster-than-light travel relates to backward time travel."

"Let me show you. Remember we talked about the Lorentz transformation a few days ago? I want to develop an interesting coordinate-axis system using spacetime diagrams. Let's start with the horizontal axis. If the x,t system is stationary with respect to the x',t' system moving with speed v in the x (or x') direction, we have:

$$x' = \frac{x - vt}{\sqrt{1 - (v/c)^2}}$$

$$t' = \frac{t - vx/c^2}{\sqrt{1 - (v/c)^2}}$$

These equations made sense for us when v was less than c." You pause. "Now, let me define a *cosmic moment line*, $t =$ constant, using spacetime diagrams." You sketch a horizontal line in the air. "It's a line paral-

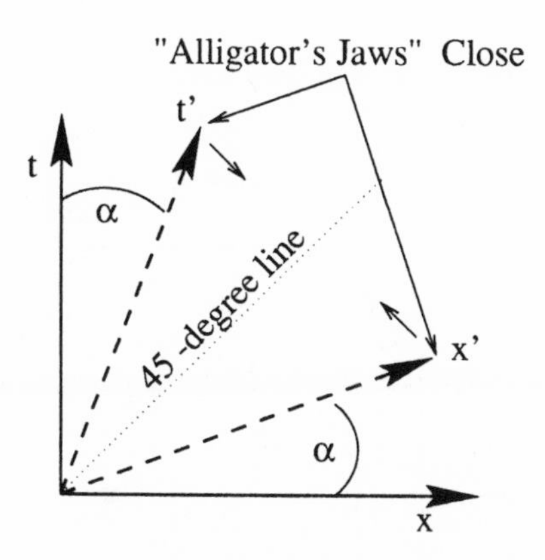

Figure 11.1 Spacetime coordinate rotation by relative motion. As the velocity of the moving system increases, the spacetime axes rotate like the closing of an alligator's jaws. At the speed of light, the alligator's jaws have clamped together along the 45-degree line.

lel to the x-axis with a fixed time coordinate. Similarly, for a moving system we would represent the cosmic moment line as t' = constant." You draw another horizontal line on a spacetime diagram.

"Now, I'm interested in a particular cosmic moment line, the $t' = 0$ line, which passes though the point $x = 0$ at $t = 0$. Using the second Lorentz equation I just listed we get: $t = vx/c^2$. We can make this more compact by using a standard convention where c is set equal to one. This gives us $t = vx$.

"Next let's develop the vertical axis of our coordinate system. A stationary particle in the x,t frame would be represented by a vertical line with equation $x =$ constant. In the moving system we have $x' =$ constant for a world line of a stationary particle in that system. Let's think about the t' axis. The t' axis is the $x' = 0$ world line of a particle stationary at the origin of the moving system, and it passes through the $x = 0$, $t = 0$ point. Using the Lorentz transformation, it has the equation $x = vt$."

Constantia takes a deep breath. "Can you summarize all this?"

"Sure. For our moving system, the vertical (t') axis is represented by the equation $x = vt$. The horizontal (x') axis is represented by the equation $t = vx$."

You stop walking, take some chalk from your pocket, and stoop down next to the sidewalk. You start to draw the two spacetime coordinate systems superimposed on one another (Fig. 11.1).

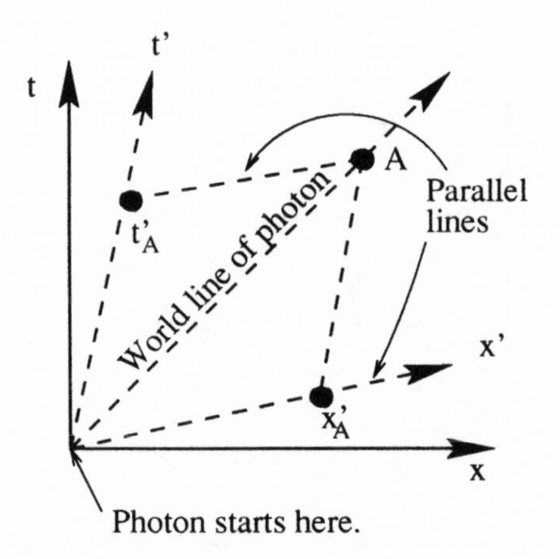

Figure 11.2 The world line of a photon is invariant.

"Why are you drawing the new system with such strange angles?"

"When the moving system is at rest, and $v = 0$, then the two systems coincide—because we have $x = 0$ for the vertical axis and $t = 0$ for the horizontal axis. As the velocity of the moving system increases, there is a strange rotation of the spacetime axes: The time axis swings down by angle α, and the space direction swings up by α. Think of the two axes as an alligator's mouth closing its jaws as the moving frame of reference goes faster ($\alpha = \tan^{-1}(v)$). The alligator's mouth is aligned at 45 degrees to the *x*-axis."

Constantia gets out a ribbon from her pocket and ties back her hair. "I see. If the moving frame is slower than the speed of light (subluminal motion where $0 \leq v \leq 1$) then the alligator never quite closes his jaws. In other words the angle α is between 0 and 45 degrees."

You nod. "When the moving system moves at the speed of light ($v = 1$), the angle is 45 degrees, the alligator's jaws have clamped together, and the x' and t' axes coincided. Time and space have united. They are indistinguishable!"

Constantia shakes here head. "How would the world lines of a photon traveling at light speed differ in the two coordinate systems?"

"They'd be identical. Both would have a slope of 1." You sketch again (Fig. 11.2). "Let a photon start at $x' = 0$ and $t' = 0$. Let it travel to point *A* with coordinates ($x' = x_A'$) and ($t' = t_A'$)."

"Sir, why did you draw slanted lines to get the coordinates for *A*?"

"Just as in the usual orthogonal x,t system, here I draw lines to form point A parallel to the x' and t' axes until they intersect the t' and x' axes, respectively. Notice that x_A' and t_A' have the same length just as they do in the unprimed system. Speed is distance traveled divided by time, so $x_A'/t_A' = 1$, which is the speed of light."

As you travel through the city, you see people dressed in monk's habits of coarse white, nuns with stiff grey wimples, princes in puffed sleeves of velvet, naked men who walked as though they had never known clothes, metallic dresses coated with glittering silks made from sea slug embryos, soldiers in olive green boots, peasant's tunics made of paper, and fine tailored wool suits. Some of the teenagers wear gowns of purple and have hair of all colors tangled and mingled with robin's eggs. Their faces are also of all colors, with stripes forming a Fibonacci sequence around their foreheads. Older people sometimes kneel with hands clamped, perhaps praying for an earlier time with less confusion and intermingling of races and species from other worlds. To your right is a woman with a bald pink head, tenderly wrinkled at the neck, with earrings the size of cattle prods.

Constantia looks at your diagram. "Do other world lines look the same in both coordinate systems?"

"No, the speed of light is the only invariant speed under the Lorentz transformation. Again, most modern physicists interested in special relativity emphasize this invariance of the speed of light, rather than the idea that the speed of light is a limiting speed." You pause. "Now I want to tell you again about something really wild—ultraluminal particles that travel so fast that they are everywhere at once!"

"Great Yggdrasill." Mr. Veil says as a robotic pigeon alights on his head. The pigeon's eyes seem to glow. Hopefully, the pigeon doesn't conceal a hidden video camera enabling the government to spy on you.

"If you travel faster than light in the x,t system, then there exists a slower-than-light x', t' frame for which a particle is infinitely fast!" You draw on the sidewalk (Fig. 11.3). "Let me draw an FTL particle's world line in the x,t system. Notice that its world line is below the ordinary photon's world line. Let's use the symbol w to represent the speed of this FTL particle. β is the angle it makes with the x-axis. Now let's move the moving frame of reference with a speed such that its x'-axis coincides with the FTL particle world line. At this point, $\alpha = \beta$. Since its space axis, x', coincides with the FTL particle world line, the particle

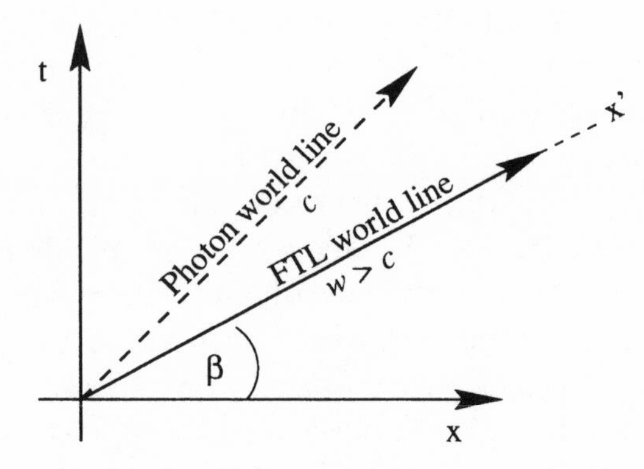

Figure 11.3 World line of a faster-than-light object.

will appear to an observer in the x', t' frame to be everywhere at once, which means the particle appears infinitely fast! If an FTL particle moves with speed w in the x,t frame, then an observer in the x', t' frame moving with subluminal speed v will see the particle as infinitely fast when $w = c^2/v$. If the FTL particle's speed w is greater than c^2/v, then we say the particle is not merely superluminal but *ultraluminal.*"

"But if a particle is infinitely fast when it travels at $w = c^2/v$, then what actually happens when it has a greater speed?"

"Let's draw a spacetime diagram, and extend the x', t' axes back to negative values." You draw on the sidewalk (Fig. 11.4). "Let's choose two events. *P* is when President Kennedy was elected president. *S* is when he was shot by Oswald. Next let's plot their spacetime coordinates in both the x,t and x', t'. For the unmoving x,t frame, we see that the time of the shooting t_s is greater than the time when he was elected President t_p. In other words, the ultraluminal particle is moving forward in time from *P* to *S* and is moving in space along the increasing *x*-axis."

Constantia's eyes light up. "I see where this is heading. In the moving x', t' system the time of Kennedy's shooting t_s' is less than the time of his election t_p'! The time order is reversed for an observer in the x', t' frame."

Mr. Veil slaps his forelimbs together. "Great Yggdrasill! To the moving observer, the particle seems to be traveling backward in time."

"Hold on! There's something else you should know! The energy of the particle goes from positive to negative when the particle goes ultraluminal."

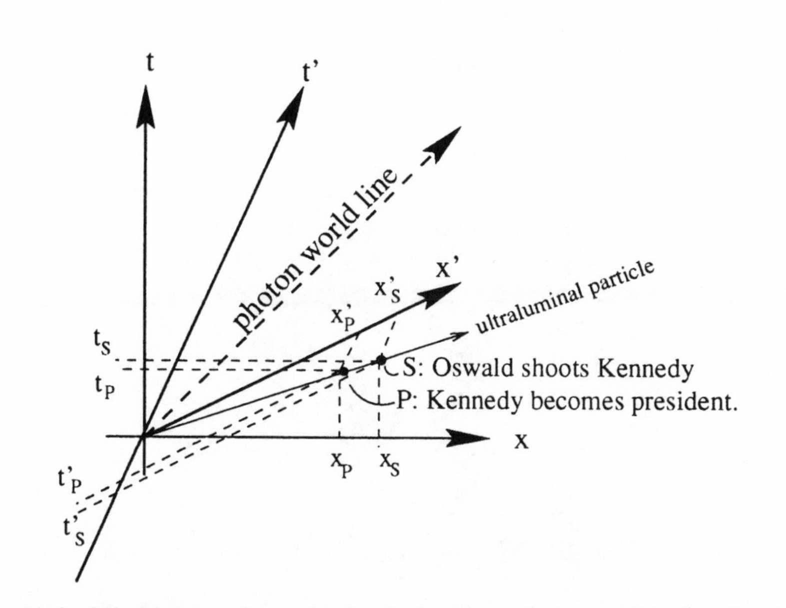

Figure 11.4 World line of an ultraluminal object. Is it possible for President Kennedy to be assassinated *before* he is elected President?

Mr. Veil comes closer. "Explain."

"If the energy of the particle in this stationary system is *E*, then the energy in the moving system is:

$$E' = \frac{1}{\sqrt{1 - (v/c)^2}} \quad \frac{\mu(c^2 - wv)}{\sqrt{(w/c)^2 - 1}}$$

Recall that I said that the particle goes ultraluminal when its speed *w* exceeds c^2/v, and it's at this point that it appears to go backward in time. If you plug values into the E' formula, you'll see that at this point the equations produce negative results. Negative energy moving backward in time in one system is positive energy moving forward in time in another."

Constantia slaps her hands together. "This negative energy is confusing to me."

"Right, historically this has led to what's called the *reinterpretation principle* (*RP*). Let's assume an ultraluminal particle is emitted from a source *S1*. It travels to an absorber *S2* ($S_1 \rightarrow S_2$). *S1* and *S2* are in the same reference frame, and an observer in that frame considers the par-

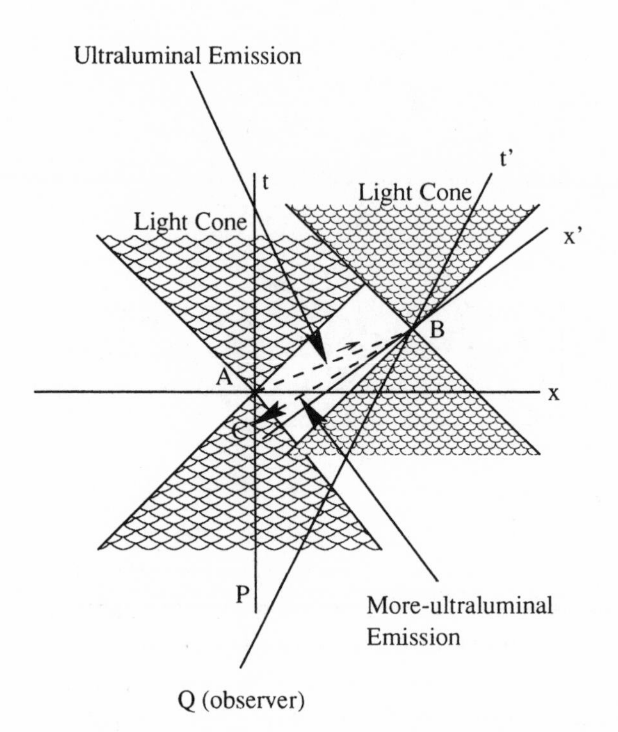

Figure 11.5 An FTL causal paradox, based on the ideas of Shoichi Yoshikawa.

ticle energy E as positive. On the other hand, it is possible to find another observer in a relatively moving frame for whom this particle would appear as a negative energy particle moving backward in time. For this observer, emission by *S1* of negative energy *increases* the energy of *S1*, and the absorption of negative energy by *S2 decreases* the energy of *S2*. The decrease of *S2* for the moving observer occurs before the particle is emitted at *S1*. The moving observer therefore would naturally interpret this process as the emission of positive energy by *S2* followed by absorption at *S1*. In other words, the RP flips the roles of transmitter and receiver. This reinterpretation can avoid many of the problems advanced against backward time travel, and physicists use the RP to eliminate various paradoxes."

"*Many* of the paradoxes?"

"Not all. For example, there are some cases where the RP seems to create lots of causal paradoxes. Consider an ultraluminal particle emitted by an observer P located at A at time $t = 0$ in the stationary system."

You draw on the sidewalk (Fig. 11.5). "A relatively moving observer (*Q*) receives the particle at *B*. Instead of *Q* seeing a negative energy particle absorbed at *B*, because of the RP he interprets it as if it were a positive energy particle emitted at *B* and traveling back down the x'-axis in the negative direction. Now, what happens if *Q* emits a second tachyon that goes even faster down the negative x'-axis the moment *Q* observers the emission at *B*?"

Constantia looks into the sky, as if for divine guidance. "This new tachyon is even more ultraluminal than the one originally emitted at *A*?"

"Yes, that's why I've drawn it below the first world line. The new tachyon is absorbed at *C*, and even observed by a past version of *P*, since *C* occurs earlier on the *t*-axis than does *A*. Again, because of the RP, *P* sees this second tachyon as a negative energy particle and interprets it as the emission of a positive energy particle."

"Weird. This means that the emission of this tachyon at $t < 0$ has been caused by emission of the original tachyon at $t = 0$."

"Right, the RP seems to have forced us into backward causation with P seeing something happen at $t < 0$ because of something he will do at $t = 0$."

Constantia puts her hands on her hips. "But what if after P sees the $t < 0$ event, he chooses not to emit the original tachyon?"

"It seems that tachyon communication systems transmitting messages from the future to the past upon request lead to problems—"

You are minutes east of the Empire State Building when you take a short cut in an alley toward a trans-species coffee shop. You look around the alley. There are two Orientals sitting in the corner: a man in his twenties with a beard and tie-dyed T-shirt, and a girl who must be only around sixteen. Her nose ring glows, casting green shadows over her heavily made-up face and dilated catlike eyes. A few FOPs line the alley, silently staring. Is it possible that some of these apparently diseased watchers are government agents or police in disguise?

You walk another few feet. It seems the whole area is filled with figures—some human, others Zetamorphs, many of indeterminate species and sex. The closest of the humans sits on a white wicker chair—a teenager with a punk hairdo, translucent jeans, and a Ralph Lauren tee-shirt under an Elvis Presley motorcycle jacket. On each of his twelve fingers are gold rings.

You suddenly hear a voice to your left. "What are you staring at?" it says.

Standing before Constantia is a six-foot tall, pea-green creature. Its eyes glow a fierce red. From beneath its thin, drooling mouth, a hundred throat appendages quiver aperiodically. Two of its off-white snouts begin to sniff at you and Constantia.

"This place gives me the creeps," Constantia whispers to you as several unmarked black helicopters begin to circle high in the sky.

You nod. "I think we've had enough fresh air. Why don't we head home?" ✍

The Science Behind the Science Fiction

> We all agree your theory is crazy; what divides us is whether it is crazy enough to be correct.
>
> —Neils Bohr, commenting on a proposal regarding the ultimate nature of matter

Faster-Than-Light Travel

Albert Einstein's theory of relativity doesn't preclude objects from going faster than light speed; rather, it says that nothing traveling slower than the speed of light (for example, you and me) can ever travel faster than 186,000 miles per second, the speed of light in a vacuum. However, FTL objects may exist so long as they have never traveled slower than light. Using this framework of thought, we might place all things in the universe into three classes: those always traveling less than 186,000 miles per second, those traveling exactly at 186,000 miles per second (photons), and those always traveling faster than 186,000 miles a second.

In 1967, the American physicist Gerald Feinberg coined the word *tachyon* for such hypothetical FTL particles. The name comes from the Greek word *tachys* for "fast." In contrast, *tardyons* are the slower-than-light particles with which we are familiar (e.g., protons and electrons). Sometimes tardyons are known as *ittyons* from the Hebrew for "slow." Tardyons have mass, but the very light ones are relatively easy to accelerate to near light speed. For example, the electrons that produce an image on a television screen travel at about 30 percent of Einstein's limit when they hit the phosphor screen. Electrons in Stanford's linear accelerator can be made to lag behind light speeds by only a few parts per billion, less than one mile per hour.[1]

Aside from the tardyons, there are also the massless *luxons* that travel *only* at the velocity of light. Luxons include photons, hypothetical gravitons, and possibly the neutrino. Because they are just lower-frequency versions of visible light, television waves, radio, and radar also travel at light speed as does electromagnetic radiation of higher frequency such as ultraviolet light, X rays, and gamma radiation. (By the way, light travels almost a million times faster than sound.)

Why does it seem that objects cannot start at a speed less than light and go faster than the speed of light (and hence make time go backward)? For one thing, special relativity states that an object's mass would become infinite in the process. Thus the Star Trekian idea of starting at a sublight speed and going faster than light contradicts the special theory of relativity. (Tachyons don't produce this contradiction because they never existed at sublight speeds. They do, however, travel backward in time.)

We know for sure that as an object's speed increases, a particle becomes heavier and more resistant to further acceleration. This relativistic mass increase is a well-tested phenomenon for high-energy physicists. In laboratories all over the world, the mass increase of elementary particles as they approach the light barrier is well known. However, physicists can imagine bypassing the mass-increase barrier not by accelerating a particle to light speed and beyond, but by putting enough energy together in one place to create a particle, like a tachyon, that is born traveling faster than light.

There is another reason why relativity seems to preclude FTL travel. If you start out moving slower than the speed of light and go faster and faster, time runs more and more slowly until, at the speed of light itself, it comes to a stop relative to stationary observers. You can't go any faster because the speed of light is an impenetrable barrier. In a strange way, if you try to increase your speed there is no time left in which to make the increase. However, while special relativity seems to forbid backward time travel, the general theory of relativity (which includes gravitational effects) may permit it, as we'll see in later chapters. Remember, special relativity describes how objects move far away from massive objects like stars while the general theory of relativity is more powerful and capable of describing spaceships accelerating near stars and black holes. Note, however, that although Einstein's general relativity allows for some forms of time travel, the energies necessary to twist time into a "circle" are so great that Einstein's equations may break down as quantum theory takes over. This means that significant further research is required before final conclusions can really be made.

If time-traveling tachyons are discovered, or if other distortions in spacetime can ever be used for time travel, then the principle of causality, that cause precedes effects, is discarded, and the chief theoretical obstacle to time travel has been removed. (This obstacle, sometimes called the "causal ordering postulate," forbids all spacelike causal connections.) Some physicists have suggested that many tachyons were created at the moment the Big Bang created our universe. However, in minutes these tachyons would have plunged backward in time to the universe's origin and been lost again in its primordial chaos. Possibly they would become isolated from our universe of tardyons and create their universe that is forever separated from us by the speed-of-light barrier.

Tachyonic Aliens

Theoretically, we lazy tardyons can approach the speed of light but will never reach it until infinite time has elapsed (Fig. 11.6). FTL tachyons, which

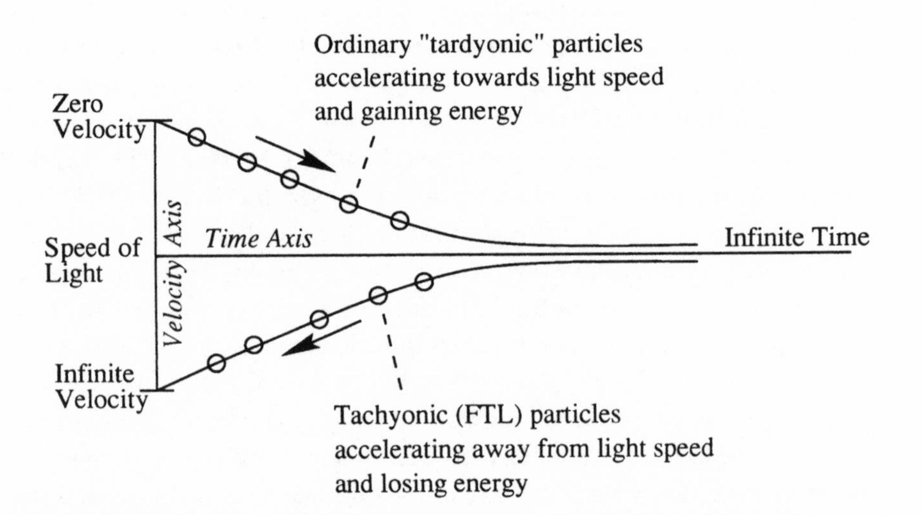

Figure 11.6 Ordinary tardyonic matter can never accelerate sufficiently to reach the speed of light. Similarly, tachyons can never slow down sufficiently to reach it. A tachyon universe would be a mathematical mirror of our own. We still do not know whether tachyons can exist in our universe.

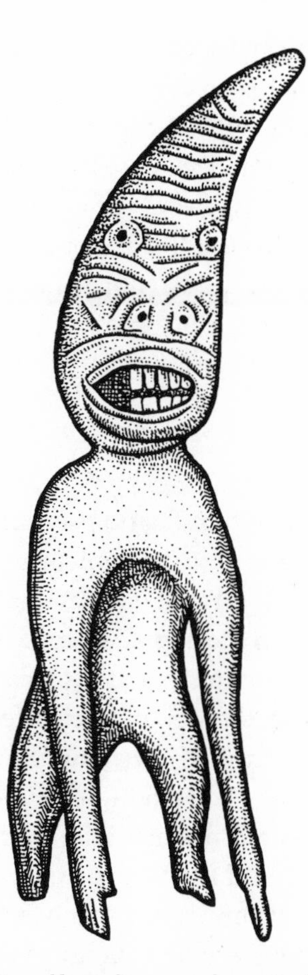

Figure 11.7 A tachyonic alien. If an alien, made entirely of tachyons, came toward you from his spaceship, you would see him arrive before you saw him leave his ship. The image of him leaving his ship would take longer to reach you than his actual FTL body. The creature would also visually appear to be traveling away from you. This means you would see a tachyonic alien as receding backward in time!

can exist in theory, would need infinite energy and infinite time to slow down to the speed of light.

If an alien made entirely of tachyons came toward you from his spaceship, you would see him arrive at your doorstep first, and then see him leave his ship. This seems strange, but the image of him leaving his ship would take longer to reach you than his actual FTL body. Stranger yet, the creature (Fig. 11.7) would visually appear to be traveling away from you, back to his ship.

In short, you would see a tachyonic alien as receding backward in time! This is not some optical trick—it is an actual fact predicted from the theory of relativity. These ideas are usually discarded in science-fiction stories. For example, if Captain Kirk in *Star Trek* were looking out the Enterprise's window in search of a ship coming toward the Enterprise at warp (FTL) speed, he wouldn't actually see the object until after it had arrived! Captain Kirk's universe would be filled with ghost images of spaceships that long ago arrived where they were going at warp speed.

Tachyonic Robin Hood

Let's consider some additional ramifications of tachyonic matter. If a tachyonic Robin Hood shot an arrow at an apple, we would see the hole in the apple before we saw Robin Hood pull back on his bow. In fact, once we had information about the hole in the apple, we could send a tachyonic message to Robin Hood forbidding him to shoot, thus creating a time paradox.

If tachyons are being created today, physicists feel they might detect them in cosmic ray showers. Additionally, they might be detected in records of particle collisions in the lab. For example, hundreds of cloud chamber records of collisions between fundamental particles have been scanned by physicists for tachyons. To find the tachyons, physicists look for reactions with more mass and energy coming out than went in, which would suggest that tachyonic interactions were giving energy to the system. To date, no evidence of this kind has been found.

There have been additional experiments to find tachyons, such as the unsuccessful attempt to create electrically charged tachyon pairs by bombarding lead with gamma rays. As the lead is bombarded, physicists looked for electromagnetic radiation emitted by an electrical charge exceeding the speed of light in the local medium. This kind of radiation is called Cerenkov radiation. Although no particle has been observed to travel faster than the speed of light in a vacuum, particles can travel in a material medium faster than the speed of light in that medium.

Let me clarify Cerenkov radiation with an example. Light in most transparent materials travels slower than in a vacuum. In these materials, tardyons are allowed by Einstein's relativity to exceed light speed in the medium. This kind of FTL behavior is allowed because Einstein's speed limit is the speed of light in vacuum, not the speed of light in glass or plastic. In 1934, Russian physicist Pavel Cerenkov discovered that whenever an electrically charged particle travels faster than light in a material, the particle gives off light. This emission, now called Cerenkov radiation, is analogous to a jet plane's sonic boom when it exceeds the speed of sound. Cerenkov radiation is a kind of "optic boom." Since tachyons always travel faster than the speed of light in a vacuum, a charged tachyon should emit Cerenkov radiation. Charged tachyons should glow as they lose all their energy and enter the zero energy/infinite velocity state called a "transcendent" state.

By searching for luminous tracks of flashes of light in seemingly empty space, researchers may someday discover the presence of transcendent charged tachyons. If tachyons are uncharged, they may be detected in other ways, but to date no method has revealed their presence. A natural source of tachyons could include cosmic rays, intensely energetic particles of uncertain origin, that continually bombard Earth. One way to search for tachyons in cosmic-ray air showers is to examine detector records and look for anomalous events that precede the main shower front.

Imaginary Rest Mass

As we've discussed, a tachyon possesses an imaginary rest mass; that is, the square of its mass is a negative number. As a tachyon increases its speed, it loses energy. Once a tachyon has lost all its energy, it travels at infinite velocity and simultaneously occupies every point along its trajectory. Particles that live in this bizarre state of omnipresence—zero energy and infinite velocity—are called "transcendent." If we wanted to reduce the speed of a tachyon, we must add energy. To slow a tachyon down to light speed requires an infinite quantity of energy. Therefore, the speed of light is a lower limit to a tachyon's velocity. Those physicists who assert that tachyons are unphysical entities do so, in part, because of a tachyon's imaginary rest mass. Supporters of tachyons counter that, since a tachyon can never be brought to rest, the tachyon rest mass is not experimentally meaningful and does not need to be represented by a real number. They also counter that a pleasing symmetry exists between tardyons and tachyons, each of which stays on its own side of the speed-of-light barrier.

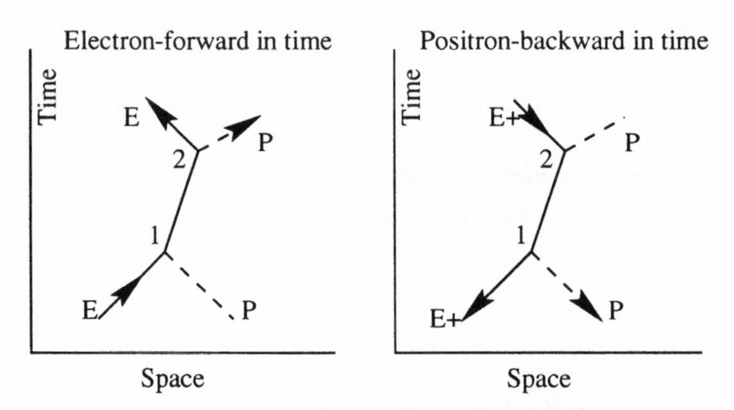

Figure 11.8 Particles in time. (*Left*): An electron (E) collides with a photon (P) at point 1, absorbs it, changes direction, emits a photon again at 2 and changes direction. (*Right*): The same particle track can be interpreted as showing a positron (E+) traveling back in time.

Richard Feynman and Time Travel

A positron, also called a positive electron, is the antiparticle of the electron. It is a positively charged subatomic particle having the same mass and magnitude of charge as the electron. Positrons were the first of the antiparticles to be predicted and discovered. P. A. M. Dirac postulated their existence in 1930 and Carl David Anderson established their existence in 1932 while studying cloud-chamber photographs of cosmic rays. Stable in a vacuum, positrons quickly react with the electrons of ordinary matter by annihilation to produce gamma radiation. Positrons are emitted in certain radioactive nuclei and are formed in pair production, in which the energy of a gamma ray in the field of a nucleus is converted into an electron–positron pair.[2]

Some physicists find it convenient to view the positron as an electron traveling backward in time (Fig. 11.8). In fact, in 1965, physicist Richard Feynman received a Nobel prize for his application of spacetime concepts to quantum mechanics in which antiparticles are viewed as particles momentarily moving into the past. Few scientists really believe that positrons are electrons traveling backward in time. This idea is regarded as a metaphor rather than a description of reality. Nevertheless, the laws of physics say that a positron is indistinguishable from an electron traveling backward in time.

All this talk about particles and antiparticles reminds me that if time travel were possible, it would be easy to be in two places at the same time. Let me explain why. Imagine that you are taking a trip from New York City to Los Angles. Once you reach the halfway point, assume that time flows backward, so you reach Los Angeles at exactly the same time you started your trip. You would be in two places at the same time. One can think of your journey in the framework of particle and antiparticle, without ever considering the notion of a backward flow of time. Consider an example. Bill Clinton is in the White House and wishes to travel to Arkansas. He and his antimatter twin (Notnilc Llib) start their trip, one in the White House, and one in Arkansas. They travel toward each other, and meet at the mid-point of their trip. This is analogous to Bill Clinton traveling from the White House to the mid-point and then continuing his journey to Arkansas with backward time flow. If Bill Clinton and Notnilc Llib met, the two would annihilate each other, producing energy. We wouldn't need to refer to time flowing backward because Notnilc Llib traveling from Arkansas to the mid-point in forward time is actually Bill Clinton going from the mid-point to Arkansas in backward time!

Imaginary Time

We've discussed $\sqrt{1 - v^2/c^2}$ as the factor (sometimes called tau), by which mass, length, and time are altered. When v, the velocity at which you are traveling, is much smaller than c, the speed of light in vacuum, tau is almost equal to 1, and common-sense physics works as usual. What happens to tau if you try to travel faster than light? Ignoring the practical problem of how to reach FTL speeds without at some point traveling at the forbidden speed c, let us look at the mathematics. At FTL speeds, the square root in the formula becomes negative. Square roots of negative numbers are called imaginary numbers, because there is no real number that gives a negative number when multiplied by itself. The values for mass, length, and time flow for your FTL body have all become imaginary numbers. There are two ways of looking at this result. First, you might use it to suggest that FTL travel is not possible. The second way is to try to physically interpret the imaginary numbers. After all, imaginary numbers appear in all sorts of practical physics problems. Even though imaginary time may make little sense in our universe, perhaps it can make sense in a universe where all objects are on the other side of the light barrier. As we discussed, in this other universe, all objects move faster than light, and it would be impossible to slow them down to c, because you would have to use energy to slow down an object, and the required energy would grow to infinity as the speed *dropped* to c.[3]

Science Fiction and Causality

Science-fiction authors have often used tachyons to achieve FTL travel or for information exchange. For example, in Bob Shaw's *The Palace of Eternity,* a million-ton tachyonic spaceship travels at 30,000 times the speed of light! In Gregory Benford's novel *Timescape,* future humans use tachyonic messages to warn the past about ways to avoid severe ecological damage to Earth. Even though we can't convert tardyons to tachyons, it might be possible to generate tachyons, send them out, and modulate them to transmit information. This would be a means for sending information to the past. If there were an entire universe made of tachyons, it would appear to inhabitants of this universe that our own universe was where FTL travel is possible—due to a symmetry in the mathematics. So there would be no travel advantages in this universe.

As we've discussed, tachyons are likely to behave as though they travel backward in time and thereby destroy the rules of causality. Causality simply says that effects happen after causes. It's an empirical law—one that has never been proved to rule the universe; however, causality does lie at the root of humankind's science. A piano will make no sound until someone strikes a key. People cannot die from an atomic bomb explosion until the bomb is detonated. These examples may seem obvious, but causality violation is used by many physicists to prove that FTL journeys are forbidden. Still, there may be exceptions to the rules of causality, especially when we are in the realm of black holes or quantum theory of subatomic particles.

Ordinary high-speed travel does not violate causality. Although relativity theory allows observers moving at different speeds to see the same events occurring in different sequences or at different times, no observers, however they are moving, will ever see an arrow hit its target just before the arrow is fired. Remember the experiment Mr. Veil conducted in the "relativity of simultaneity" with a pile of newspapers and magazines? Different observers may disagree about whether the laser beam reaches the two ends of the ship at the same time, or if the magazines are hit before the newspapers; but all observers agree that the laser blasts leave the lasers before they arrive at the newspapers and magazines.

Time Travel by Ba

Since
the era of H. G. Wells,
science fiction buffs have
gone gaga over the idea of
traveling through time.

—David Freedman, *Discover,*
1992

The advantage of
being a theoretical physicist
is that you never have to
worry about the cost of a
thought experiment.

—Yakir Aharonov

168

I cannot change the laws of
physics, Captain.

—Scotty, *Star Trek*

We could imagine a
world in which causality does not
lead to a consistent order of earlier and
later. In such a world the past and the
future would not be irrevocably separated,
but could come together in the same present,
and we could meet our former selves of several years ago and talk to them. However, it
is an empirical fact that our world is not
of this type. Time order reflects the
causal order of the universe.

—Hans Reichenbach, *The Rise of
Scientific Philosophy,* 1951

Time Travel by Balloons and Strings

12

✍ You hear a screaming sound tear across the sky, a horrifying cry, a sound that shatters the relative quiet of Mercer Street like the screeching of fingernails upon a blackboard. "What the hell is that?" you yell, looking upward.

Constantia stops walking. "My God," she shouts. She cups her hands over her eyes to help her see into the bright sky.

There is a flash of brown as the hawk crashes into some parked cars—a flurry of beating wings, sharp claws, raucous screeching. Startled, you jump to the left as your heart thumps like an anxious conga-drum player. Mr. Veil also veers left, an unpracticed action that nearly makes him topple over.

"Nothing to worry about," Mr. Veil shouts looking all around for traces of the bird. "Just a hawk."

You attempt to will your heart to shift into neutral. "Mr. Veil, what was wrong with it?"

"Probably hit a plane. One of its wings looked cut."

Constantia bumps into you. "Don't see any planes around here."

You look up and see dozens of hawks gliding

like windblown bits of ashes high above the buildings. Others seem to be joining them from Varick Street. What are they looking for? The decaying remains of a road kill? Pigeons?

"Let's keep going," you say as you walk with Mr. Veil and Constantia toward Greenwich Village along Mercer Street.

After a few minutes, Constantia turns to you. "I'm starved," she says. She is wearing a pretty, loose, old-fashioned dress with a fractal pattern and long, white, lace-trimmed sleeves. Her dark hair is pulled back by a turquoise velvet bow, and her eyes are the most beautiful eyes you've ever seen outside of a fashion magazine. Large eyes, with long lashes.

"Okay," you say.

When you get to Washington Square, the three of you grab some sushi at a mom-and-pop Japanese joint. You enjoy the Ganymedean squid as Constantia and Mr. Veil slurp their economy-sized bowls of miso soup, flavored with delicate great scallions and a dash of ginger. After eating, you take the subway express to New Christopher Park. New York City trains are efficient, clean, and—given the astronomical cab fares near the village—cheap. You pay 2000 yen per subway token—just over $20.

You wander the five-train car once during the two-minute trip, but mainly you sit, nose pressed to the window, trying to look at the fashion advertisements as your train passes in front of them at one hundred miles an hour. For some reason, brain-convolution patterns are the rage for women's wear. You reach the park within twenty seconds of the arrival advertised in the printed timetable.

You turn to Constantia and Mr. Veil as you step off the train. "In our last few lessons, I'm going to try to avoid almost all mathematics and tell you about various ways to time travel: time balloons, cosmic strings, wormholes, Gödelian universes, and Tipler cylinders. Today we'll talk about time balloons. To use these for time travel, we'd need only one piece of exotic equipment: a massive balloon capable of being rapidly shrunk or expanded to any of a wide variety of sizes. If you had such a balloon, you could step inside and be transported to any time in the future or past."

Mr. Veil brings out a pipe and begins to puff on it. "Sir, why would inflating or deflating the balloon move a person in time?"

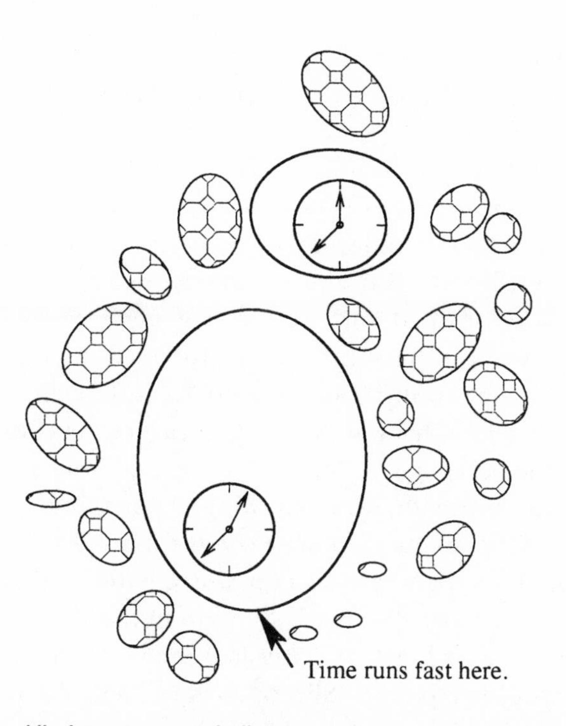

Figure 12.1 All objects—even balloons—exert gravity. In the expanded balloon in the foreground, gravity is "diluted" over a larger area. In a contracted balloon, gravity is concentrated. Since gravity slows down time, clocks inside the different balloons run at different speeds.

"Do you recall when we talked about gravity slowing time near a black hole?"

Mr. Veil nods. "Yes, we even said that time travels more slowly near the ground on Earth than away from the ground, for example, when you're in a plane."

"Yes, all objects exert gravity. Pretend I were to put you in such a balloon and blow it up. The balloon, as it gets bigger, exposes you to an increasingly reduced gravitational drag on time because the balloon would be exerting the same gravitational effect over a larger volume, diluting its strength. For you, time would speed up inside the balloon."

"Does this mean that if the balloon contracted, its gravity per unit volume would increase, and time would slow down for me inside the balloon?"

"Right." You hand Mr. Veil a huge balloon (Fig. 12.1).

"Please step into this."

"Is this safe?"

"Yes."

Mr. Veil steps into the balloon. He looks like a little lost fish in the bottom of a fishbowl. You turn to Constantia. "Constantia, will you please blow up the balloon?"

Constantia eyes you hesitantly, steps forward, places the balloon's nipple in her mouth, and begins to inflate the balloon. Inside, Mr. Veil looks at his wristwatch to measure the effects of the balloon on time warpage. After five minutes, Constantia stops blowing and, out of breath, says, "Nothing's happening."

"Exactly. The problem is that the effects are immeasurably small, and you'd need an extremely massive balloon that rapidly changed size if you wanted to use it to jump in time."

Constantia clenches her fists. "You had me blow that up knowing nothing would happen?" She takes her suede boot off and throws it at you.

You artfully dodge the boot. "Constantia, I wanted to make a point so you'd remember it. In any case, even with a massive balloon it wouldn't be standard time travel because the balloon would make Mr. Veil younger or older, rather than have him travel through time."

"Sir," Mr. Veil says from within the balloon. "You mean I wouldn't be sent into the world's past or future but I'd be sent into my own personal past or future?"

"Yes. There are some dangers. Go too far and you could be translated into a decaying corpse. However, you personally would feel as if you had experienced your normal life span inside the balloon."

Constantia looks at Veil through the wall of the balloon, as a few passersby stop to gawk at the three of you. A little boy in a boy-scout uniform starts jumping on the balloon.

"What if the balloon made him too young?" Constantia says. "What if it transported him to a time before his birth? His mother wouldn't even be around."

Mr. Veil is gasping for breath inside the balloon. "Please leave my mother out of it."

A pigeon alights on the balloon. You ignore it and turn to Constantia. "I'm not sure. Perhaps we'd see Mr. Veil break up into atoms." You pause. "That's enough for this demonstration. Let's head into New Christopher Park."

After a few blocks, the three of you start to walk deeper into the park together, your feet crunching on twigs as you wander through leafy groves, pushing aside wet brown branches—the looming skyscrapers of midtown never quite out of sight. All along the path are vines, palmettos, slash pine, bay trees, and water oaks. Constantia seems particularly impressed with the wild magnolias and cypress trees.

It is amazing what a little genetic engineering can do to help the more tropical species survive the New York winters. But perhaps the park custodians have gone a bit too far: Everywhere there is a riotous, cumbrous mass of both undergrowth and trees. There are even alligators and dark ponds teeming with mystery and dangerous ambiguities. Luckily you know where the dangers are and can avoid them.

A stringy vine hangs down by your head, and you give it a tug. You turn back to Constantia and Mr. Veil. "Another way to time travel is by using cosmic strings."

Constantia grabs the vine from your hand. "Cosmic strings?"

"They're theoretical at this point, but they may exist as long, thin bundles of energy left over from the Big Bang. They'd be so dense that a single inch would weigh 40 million billion tons—that would be like cramming 40 million billion elephants into a space the size of a strawberry."

Mr. Veil nods. "That great a density should really warp space and time according to general relativity."

"Yes, they can create shortcuts in space. Think of space as a rope that you can bend. If you placed an ant on the rope, you could make it much easier for the ant to get from one end to the other simply by knotting up the rope. Similarly, you could travel close to a string and outrace a faster object traveling a different path—even if the faster object was a ray of light. As we discussed, an object

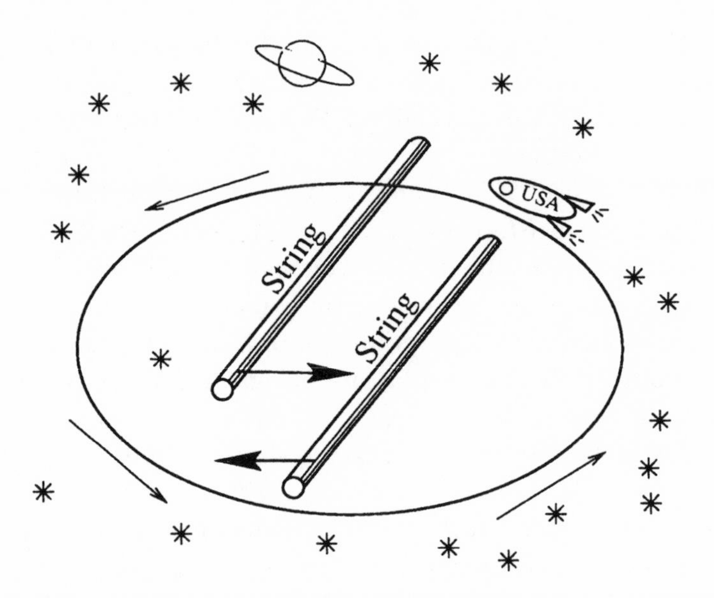

Figure 12.2 Two cosmic strings rush past each other creating a distortion in time and space. The rocket ship hooks around the spacetime shortcut created by the string moving toward it, overtakes the one moving away from it, and whips back around into the past.

moving close to the speed of light would experience a slowdown of time relative to a stationary object. If an object could move at the speed of light, it would experience a freezing of time. An object moving effectively faster than light would move backward in time. Since a cosmic string would allow a space traveler a chance to outrun a ray of light, it would allow a time warp. The effect could be increased by using two strings rushing toward one other because spacetime would be distorted to an even greater degree in the vicinity of the strings. Think of the knots in a rope becoming even more knotted."

Constantia bends down to examine a flower, and turns to you. "How would it work?"

"Imagine you are in a rocket. The two strings rush past each other, with the first string moving toward your rocket ship while the second is rushing away. Your ship would race past the first string along the spacetime shortcut it created. Next your ship races after the second string, hooks around it, and heads back along its spacetime warp. You could return to where you started well within your past!" (see Fig. 12.2)

Some of the oldest oaks of New Christopher Park are nearby, and a lovely lagoon—long, serpentine, seeming endless—winds under an arched bridge. "Come," you say. "I want to go deeper into the trees." Mr. Veil and Constantia follow you as you walk into a thicket of the oldest oaks, where the grass is high and neglected, and in which not even the poorest homeless heart would seek shelter. You make your way to a small clearing among the bone-white roots and mossy earth. The breeze from the nearby lake is cool, and for a moment there seems little scent of New York. At the far side of the lake, you see a pontoon boat that is part of the park's fleet. It is seven feet wide and seventeen feet long, with pine benches running parallel on the sides, and it could accommodate up to a dozen passengers. Park employees run guided boat tours year-round, three times a day for visitors at 300 yen a head. The visitors get an hour's opportunity to glimpse the vastness and timelessness of the verdant shores.

Constantia examines some of the dark green growths, lush as a rain forest. Cypress trees soar hundreds of feet and look prehistoric. "You know your way out of this park?" she says.

"Of course."

In the distance, a genetically engineered toucan screams. Its bright bill is precisely colored with all the colors of the rainbow, with a heavy emphasis on indigo. Constantia looks up at the sky. The clouds are turning from pink to sable. You have perhaps half an hour to get back to the museum before night descends.

That night the sound of rain wakes you. On a wall, a shadow is moving. You sit up, watching, listening. A sound of thunder breaks the eerie stillness. You can hear it echo over the roof, shaking the traffic lights. From outside comes a dull, metallic thud—a chair blowing over, a shutter breaking loose. The green luminous hands of a clock by your bed reads 2:30.

Then the rain comes again, except it isn't rain. It sounds like someone running on slippery floors. Against your window, the moon throws the shadows of a woman's head and her raised arm.

"Constantia?"

You aren't sure you really hear running sounds. All sounds are swallowed up by the noise of the rain. You roll over and hit your light switch. The sudden brightness hurts your eyes.

"Who is it?"

There is no answer. You draw back your covers, and walk slowly toward the window. You feel unsteady, as if still asleep.

"Constantia?"

You hear a voice. It must be her. You run to the window and open up the blinds.

"Constantia?"

A woman is sitting outside your window, one hand pressed against the glass, her eyes screwed up against the light. It is Constantia. You stand for a moment staring at her, trying to make sense of what you are seeing.

"Open the window!" she says.

Constantia points at the bottom of the window. You hesitate, then pull it open. The rain wets your pajamas. Suddenly you feel exposed.

Constantia steps inside and quickly slides the window shut behind her.

"Sorry to scare you like this, but I didn't want them to see me." Her voice is urgent.

"Who?"

"The Time Cops—the police of paradox. The news just broke. They want to prevent all research in time travel. They're the government's new way of benevolently overseeing the welfare of all." Constantia's tone is sarcastic as she stares at you in your pajamas. Her eyes wander from your bare feet to a spot directly in the center of your chest.

You instinctively fold your arms. "Yes, I heard rumors of them. The Time Cops want to make certain that everything will happen in the past exactly as it did happen, no matter how catastrophic an event it might have been. They're afraid some Jews will try to find a way to go back and assassinate Hitler, or some UFO buffs will go back and try to find out what really happened in Roswell, New Mexico, in 1947."

Constantia drips water onto the floor. "If we want to go back in time without everyone knowing about it, I have something to give you." She clutches at something in her hand. "You wear it on your chest, and it will distort your voice so that others won't be able to make out your words. Only those who have a similarly tuned unit can understand what you are saying. Mr. Veil and I will wear our own. This way we can all talk to one another without anyone being able to make sense of what we're saying."

"What's so urgent?"

Constantia sighs. "If you continue your lectures without the scrambler, I fear that the museum will start crawling with Time Cops. You know how it is. They don't like anyone talking about time travel. Scared that too many people would leave this Earth. Scared that the great historical personages like Jesus and Einstein would be mobbed."

You nod as you fumble with the voice scrambler. "Good thinking."

Overhead you hear a police helicopter pounding the air, and a number of ambulances turning onto Fifth Avenue and speeding away toward Rockefeller Center. Then as quick and quiet as a ferret, Constantia opens the window and rushes away into the night.

The Science Behind the Science Fiction

> In all the history of mankind, there will be only one generation that will be the first to explore the solar system, one generation for which, in childhood, the planets are distant and indistinct discs moving through the night sky, and for which, in old age, the planets are places, diverse new worlds in the course of exploration.
>
> —Carl Sagan

In this chapter, the ideas for massive balloons come from the theories of physics professor Yakir Aharonov, who *Discover* magazine called one of three prominent researchers devoting a career to studying the realities of time travel. (The two other physicists are Kip Thorne and Michael Morris, who use Einstein's general theory of relativity to convert cosmic wormholes into time machines.) The balloon machine also relies on Einsteinian relativity to provide the necessary distortion of time and space.

Aharonov's main contribution to the balloon idea is the use of quantum mechanical balloons that are linked to the behavior of one or more quantum mechanical particles.[1] According to quantum mechanics, certain particles can exist in various states simultaneously until they are observed. This means that the balloon would also exist simultaneously in all its possible sizes—and the occupant of the balloon would simultaneously exist in many, slightly different, rates of time. Although it would be hard to verify the theory behind quantum mechanical balloons, Aharonov says we may construct a small, simplified version that would send particles into their own past or future. See Aharonov (1990) and Spear (1992) in the References for a precise description of these quantum mechanical balloons.

Princeton physicist Richard Gott formulated the idea of time travel using two cosmic strings moving at near light speeds in opposite directions.[2] We do not know whether cosmic strings exist, but they might be detected someday by the way they bend light from stars. They could even cause multiple images of stars.

Cosmic strings would have very unusual dimensions. They would stretch the width of the universe, and would be extremely thin, with a diameter of only 10^{-29} centimeters and packing 10^{22} grams into each centimeter of their length. In the early 1980s, Richard Gott and W. Hiscock independently discovered that cosmic strings would bend and distort spacetime in strange ways. Gott's time machine—constructed with two high-velocity, infinitely long, straight cosmic strings—uses the ability of strings to bend spacetime and to focus light rays. Gott showed that as the two strings pass, closed timelike loops encircle the strings. If our universe is infinitely large, and if cosmic strings formed after the Big Bang (as several cosmological models predict), then the universe has already created a time machine. Even if cosmic strings do not exist, it is important that Gott formulated a time machine that does not seem to violate any known laws of physics.

Of course there is much debate in this area, and researchers such as Stephen Hawking have formulated a Chronology Projection Conjecture, which proposes that the laws of physics always prevent the creation of a time machine. Hawking believes this assertion, in part, due to the fact that we have not ever seen a time traveler from the future. Various research in the scientific literature—for example, by D. Boulware, L. Friedman, H. Politzer, C. Cutler, and A. Ori—continues the debate (see References).

I am afraid I cannot convey the peculiar sensations of time travelling. They are excessively unpleasant.
—H. G. Wells, The Time Machine, 1895
Among all creatures, humans are distinguished by the extent to which they wonder about things that do not immediately affect their subsistence.
—Lewis Carroll Epstein, Relativity Visualized
258
ff
Death is just nature's way of saying, "Hey! You're not alive anymore!"
—Bull, Night Court
Spacetime has no beginning and no end. It has no door where anything can enter. How break and enter what will only bend?
—Archibald MacLeish, Reply to Mr. Wordsworth
Time is nature's way of keeping everything from happening at once.
—Anonymous

Can John F. Kennedy Be Saved?

13

✍ Constantia is on her back, in the middle of the floor, her naked heels touching the border of a Persian carpet. A light breeze ruffles sheet music scattered near her head.

"Constantia?"

She sighs, drawing an arm across her face against the sunlight. You let her sleep a while longer as you look at some photos she has recently hung on the wall. One of them shows her with her parents, probably taken about five years ago. If anything, she is more beautiful now. There is a completeness about her, an identity. The girl in the photo, smiling sweetly through amber-painted lips, her hair teased into an elaborate punk hairstyle, seems to lack all that.

"Constantia?"

Her lips move as she breathes deeply. Fast asleep. You lean across her and pick up a sheet of music that is inches from her left hand. It is Chopin's *Andante Spianato* and *Grande Polonaise in E-flat Major*, Op. 22.

Suddenly you notice a Polaroid photograph by her side. It is a photograph of you! The picture had to have been taken at the music research station in New Guinea a few years ago. Your chief interest is time travel, but you also love music and find it a nice way to make a living. You have spent most of your adult life in one tropical country or another mapping out the huge variety of musical instruments where they were prone to destruction. You've also visited many metropolitan areas where instruments were perpetually under threat by the latest electronic or computer replacements. Among the hundreds of undiscovered ancient or primitive instruments lay the basis for new, spirit-elevating sounds, a fact recognized by music companies that sponsor your research. It is important work—a race against time. In the past, you have dedicated yourself full-time to finding ancient instruments, to warding off the crass commercial music synthesizers. Everything else had taken second place. Until now.

"Hi," Constantia says squinting up at you, her dark eyes absorbing photons of light. On Constantia's shirt is a photograph of Isaac Asimov, the most prolific and popular science writer of all time. You feel a twinge of regret. You always wished you had the opportunity to meet the man.

"Want to go to the Four Seasons?" you ask. You like the place. It's like a comfortable living room—the decor's not too formal, the food's good, and the waiters don't hover.

Constantia smiles, placing the end of her tongue between hard, white teeth. "Why don't we go for a walk first? Central Park? Shall we get Mr. Veil?"

You pause for several seconds and finally say, "Okay."

The three of you hop in a cab. The cabbie is a thin old man with unkempt hair, yellow eyes, and breath that stinks of cheese. His tight tee-shirt shows a map of South Vietnam. HELL NO, WE WON'T GO, 1968, the words beneath the map read. His jet-black eyes scan Constantia quickly, passing from her lips to her hips before appearing to lose interest.

"Where we going, lady?" he asks.

"Central Park. East Side?"

"Okay." He looks up and uses the rear-view mirror to meet her eyes. "That side gives me the creeps. A 3000-yen extra fare."

Constantia nods, and in minutes are at the east side of Central Park.

"Let's move through the woods as we talk," you say. "If you don't mind walking on dirt paths."

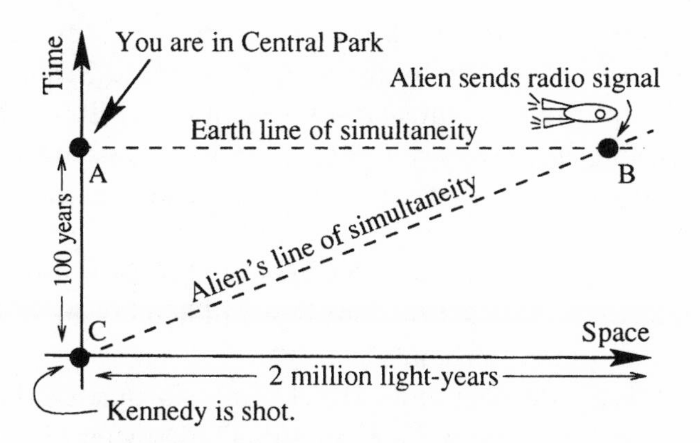

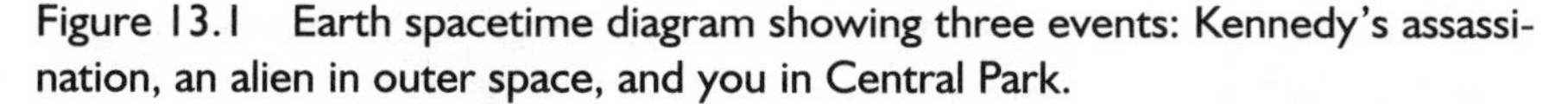
Figure 13.1 Earth spacetime diagram showing three events: Kennedy's assassination, an alien in outer space, and you in Central Park.

"Not at all," Constantia says as she brushes a little pollen from her fine spun robe, a garment that looks as if it could have been worn in the time of Cleopatra. Her entire form is slightly smaller than yours. If her hair were blonder, she would have fulfilled every promise of an angel.

"Come closer," you say. "Today I'd like to discuss some more problems dealing with simultaneity." You pause while slipping on the voice scrambler Constantia gave you the night before. Mr. Veil and Constantia already have theirs on. "John Fitzgerald Kennedy, America's youngest President, was born in 1917 near Boston, Massachusetts. He was assassinated on November 22, 1963, in Dallas, Texas, by a bullet fired by Lee Harvey Oswald. Can we use the laws of relativity to save him?"

Mr. Veil pushes aside a tree branch. "Sir, what do you mean?"

"Let's make a graph." You take out a pad of paper and begin to sketch (Fig. 13.1). "I'll label Kennedy's death at point C with a reference position $x_0 = 0$, $t_0 = 0$. Event A is us talking together in Central Park right now, 100 years after his death. Since my space axis on the graph will incorporate millions of miles, we'll give both events the spatial coordinate $x_A = 0$, and we'll consider Dallas and New York to be the same spatial point. Since today we are talking 100 years after his death, $t_A = 100$ years. Now let's assume that simultaneous with event A in our frame, an alien in a galaxy two million light years away is in a spaceship that sends a radio signal to Earth (event *B*). The alien is moving away from Earth along the straight line between the ship and Earth."

You walk slowly but steadily, and you try not to be distracted by the fern fronds, and the flowers, and the patterns on Constantina's garment. The flowers are everywhere, and it is difficult not to be seduced by them.

You reach over to a branch of a bay tree and break off some leaves. "Crush some of these leaves on your face and hands. The smell will keep away some of the mosquitos."

The bay tree is a member of the laurel family. Fortunately the leaves emit a smell that most winged creatures don't like. After Mr. Veil and Constantia spread a little of the leaves on themselves, Mr. Veil returns to your discussion of time.

"Sir, event *B* is located at $x_B = 2 \times 10^6$ light years, and $t_B = 100$ years?"

"Exactly. We know from our past discussions that simultaneity is relative, and, since the ship is moving, there is some velocity that will make its radio signal occur right around the time of Kennedy's assassination, which we will denote $x_0' = 0$, and $t_0' = 0$ for the alien. Since there is a line of simultaneity for the alien and Kennedy, is there a way that the alien can warn Kennedy, thus preventing his death and changing the course of our history?"

You come to a large spiderweb suspended over your soft path by thin, shimmering threads. Respectfully, you duck beneath it rather than destroy it. Mr. Veil and Constantia follow your lead.

You pull off more bay leaves. "Spread some more of this on your faces and arms."

Constantia looks doubtful. In contrast, Mr. Veil spreads the aromatic juice on himself with unbridled enthusiasm.

"Can Kennedy be saved?" you say. "Using these values for x, t, x', *and* t', we can employ the Lorentz equations to compute precisely how fast the alien has to travel in Earth's frame of reference in order for Kennedy's murder to be happening simultaneously in the alien's rest frame (t_B'). We can neglect the motion of the Earth around the sun, because this is so small compared with the distance of Earth from the alien. It turns out that because the distance is large (two million light-years) the speed of the alien does not have to be large to carry the alien's time back 2000 years for Earth. It also turns out that the ship must be traveling *away* from Earth to make its now the same as Kennedy's now."

"But Sir, can the alien warn Kennedy?"

"We've shown that there is a frame—the rest frame of the alien—in which Kenendy's death and the radio broadcast occur at the same time. There's no doubt about it. It would seem the aliens could warn Kennedy. However, in this frame the radio signal connecting the alien and Kennedy would have to travel at infinite speed, but this is not possible. The relation between the two events is spacelike, and spacelike events cannot have a cause and effect relation."

Constantia hangs her head low. "Even though the assassination occurs at the same time as the radio broadcast, Kennedy can't be warned." She pushes her way past a large fern, her boot crunching against its dying stem.

You come to the edge of Central Park and stand on the edge of a savanna. For the first time you catch sight of other humans—a very distant band of scantily clothed nomads moving steadily through the tall grass. Some have dyed their hair in shades of crimson, and others have painted their faces the color of red earth.

You turn to Constantia and Mr. Veil. "I'd like to give you another example why faster-than-light travel violates the normal order of cause and effect. Here's an interesting, hypothetical example. The year is 1960. John F. Kennedy is meeting with small, gray aliens on their home planet in outer space. At this meeting, Kennedy promises not to place atomic weapons into orbit. Let's mark this position as 'Promise' on our space time diagram" (see Fig. 13.2).

"Sir, how could he get into outer space? There were no interplanetary spaceships in 1960."

"Assume for the moment that the U.S. government had captured alien technology and hid the alien ship in Roswell, New Mexico. Kennedy's people kept the secret from the world, but they allowed him to travel into space to meet the aliens. After he promised the aliens not to deploy atomic weapons in outer space, he returns to Earth at 0.6 light speed." You sketch a diagonal line on your figure.

"Here's the sad part," you say. "Four years after his promise, the aliens decide that they do not like John Kennedy's decision to put a man on the moon, so they start launching faster-than-light asteroids at Earth to destroy the White House. Let's mark the time they started firing from

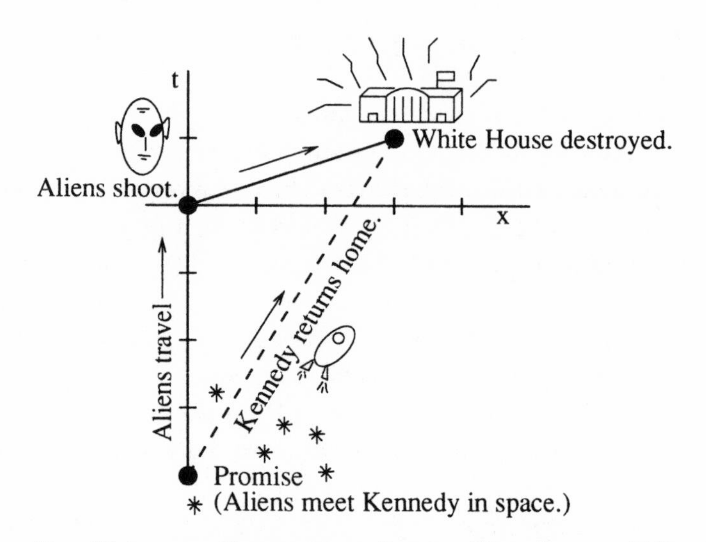

Figure 13.2 Alien ("laboratory") spacetime diagram. The alien world line is the vertical axis. At the position marked "Aliens shoot," the aliens launch asteroids at three times light speed to destroy the White House just as Kennedy arrives.

outer space position 'Aliens shoot.' Here I've drawn the path of their asteroids and show them traveling at three times the speed of light. Finally, at the event marked with a sketch of the White House, the White House is destroyed just as Kennedy arrives."

"Sir, where is this story headed?"

"I'm trying to give you another example of how faster-than-light travel seems to violate the normal order of cause and effect. Let's redraw the spacetime diagram from the point of view of Kennedy's space ship as it returns toward Earth. In this frame, the 'shoot' event can still be placed at the origin, the spacetime coordinate of 0 for rocket and stationary frames. Using the inverse Lorentz transformation, we can figure out that the 'White House destroyed' event is located at $t' = -1$ years and $x' = 3$ years, as if the firing path were reflected about the space axis" (see Fig. 13.3). "The coordinates of the 'Promise' event are located even further down on the time axis at $t' = -5$ years and $x' = +3$ years. In Kennedy's spaceship frame, the world line of the spaceship extends vertically from the promise to the White House explosion. The world line of the aliens extends from promise to the 'shoot' event."

You take a rest, and look around you. Towering over you are immense cyad trees. Never have you seen trees of this size, their monstrous leaves bigger than your entire body.

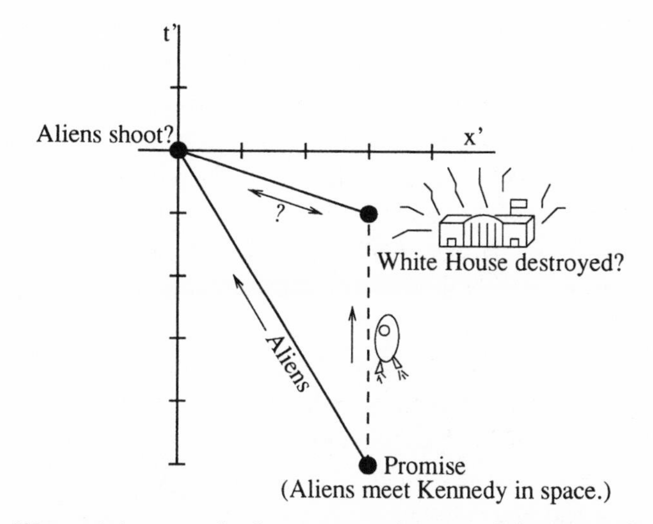

Figure 13.3 "Kennedy's space ship" spacetime diagram of departing aliens. In this diagram, the White House's destruction comes before the aliens fire.

You continue to lead Constantia and Mr. Veil, nearly oblivious to the sounds of insects and the slithering of amphibians near your feet. There are muted roars in the distance.

You are suddenly aware of a cool breeze, and you glance back. Had humans ever walked this precise path? You easily move aside the dense foliage with your left arm as you continue to walk.

After a few minutes, you come to a thick waterfall descending into a bubbling pond. You stand there, refreshed by the mist. Constantia seems to enjoy the rushing sounds of the fall. She removes her sandals and lets her feet slip into the water. You watch it swirl around her toes. Her perfectly trimmed toenails oscillate with gold and silver sparkles.

Mr. Veil grabs the pad of paper containing your diagrams. "Sir, I see something odd. In the Kennedy spacetime diagram, the projectile world line tilts downward to the right, and the destruction of the White House occurs before the scoundrelous aliens launched the projectile at the 'shoot' event. It looks like the asteroid moves with three times light speed *from* the White House *toward* the alien ship. Yet humans have certainly not created the means to do this." He pauses. "In fact, the White House is destroyed by the asteroid at the moment the White House appears to launch the projectiles."

Constantia says, "Right, and even though the spacetime diagram would make it appear that the projectiles are sent toward the aliens, the aliens did not get injured by the asteroids from the tremendous 'impact' at the 'shoot' event. What does this all mean?"

You sit down upon a log. "Let me explain. By having a material object travel faster than light in a vacuum, we produce a confusion of cause and effect. This is why some scientists believe that no signal and no object can travel faster than light in a vacuum. If this were possible, the entire network of cause and effect would be destroyed. Science, history, and philosophy would not be possible." You hesitate for a moment.

"But—?" Constantia says.

"This view is limited, because it applies to flat spacetime. Spacetime curvatures *do* seem to permit causality violations in certain cases and permit time travel. We'll use this to travel back to see Chopin."

Constantia looks at you, and you feel the pressure of her dark gaze with an agreeable soft jolt.

You walk on. The woods shift, mammoth trees giving way to slender saplings. Weeping willows, perhaps. Patches of light move over the wavy grass like lemmings scurrying to an unheard rhythm.

You peer into a valley. You can see distant skyscrapers, green slopes, and a ragged and rambling Hudson river breaking here and there for peninsulas of wild grain. The woods seem to creep up into the skyscrapers, sending their roots deep into the bricks and steel. As you travel closer to the valley, through the branches, you see the glitter and twinkling of the Hudson.

In minutes, you walk though tall grasses—wild wheat, perhaps—to the edge of the Hudson River, where it laps gently with a strong tide. Occasionally the river pulls back, exposing an extraordinary array of mussels, bottle caps, and rainbow-colored condoms.

You walk north, passing fishermen near a small town on the edge of the river. Other people are tending sheep or goats, or driving small herds of robots toward nearby settlements or walled enclosures.

"I'm hungry," Constantia says.

"How hungry are you?" you say in a loud sing-song chant, giving her the lead-in for a joke.

"I'm so hungry, I could eat the north end of a southbound Akademgorodokian porcupine," she replies.

"What's a porcupine?" Mr. Veil says.

You shake your head. "Let's go find a place to eat."

It is nighttime, and the sky deepens from purple to maroon. You look up. You can barely perceive the incredible lamp of stars through the hazy New York air. The weather patterns must have cleared an opening in the clouds. The more recognizable constellations are unmistakable. Orion's square shoulders and feet, the beautiful zigzagging Cassiopeia, the enigmatic Pleiades. You even see Aldebaran, the red star in the constellation Taurus. You remember the time you first saw Aldebaran as a young boy vacationing with your family in Ajaccio, a town in Corsica, the birthplace of Napoleon.

You walk with Mr. Veil and Constantia for a while, enjoying the surroundings. A wind is blowing, but from where you cannot tell. All of the great metropolis seems alive.

The Science Behind the Science Fiction

> Science involves myths that are of a special type, namely myths that are predictive, empirically testable (good theories always make predictions of claims that are testable by experiment) and cumulative (a good theory always encompasses all observational evidence of its predecessor and still manages to add something new).
>
> —John Casti, *Alternate Realities*

In this chapter we have emphasized that FTL travel of objects seems to violate the normal order of cause and effect. Nevertheless, various hints on the possibility of FTL links between quantum particles, as discussed in Chapter 5, have given rise to an amazing array of amateur research efforts to design superluminal communicators based on quantum connections. Many of these researchers have adopted colorful names for their "institutes," including: the Seattle Center for Superluminal Science (John Cramer); the Lompico (California) Institute for Superluminal Applications ("Changing yesterday today for a better tomorrow"—Nic Harvard); the Ansible Foundation, Davis California (Wil Iley and Mark Merner); the Galois Institute for Mathematical Physics, San Francisco (Jack Sarafatti); and the Notional Science Foundation, Boulder Creek, California (Nick Herbert). In addition, university physicists from all over the world are investigating the possible FTL quantum connection as a signaling medium.

For those of you who like solving spacetime problems of the kinds in this chapter, Edwin Taylor and John Wheeler's book *Spacetime Physics* (W. H. Freeman) is a real treat. Their text emphasizes the unity of spacetime and has become a standard for modern physics and relativity courses.

Closed Timelike Curves

If one could travel in time, what wish could not be answered? All the treasures of the past would fall to one man with a submachine gun. Cleopatra and Helen of Troy might share his bed, if bribed with a trunkful of modern cosmetics.

—Larry Niven, *All the Myriad Ways*

The most convincing argument against time travel is the remarkable scarcity of time travelers.

—Arthur C. Clarke, *Profiles of the Future*

Where beauty has no ebb, decay no flood, but joy is wisdom, time an endless song.

—William B. Yeats, Irish poet

The end of all our explorations will be to come back to where we began and discover the place for the first time.

—T. S. Elliot

Closed Timelike Curves in a Gödelian Universe

14

Damn. Zetamorphs have no sense of humor. Zero. Zippo. None. How could they? They consume great deals of hentriacontane, a hydrocarbon of the paraffin series $CH_3(CH_2)_{29}CH_3$ present in petroleum and natural waxes.

You motion to Mr. Veil's latest dish. "Mr. Veil, what is that stuff?"

As the three of you dine together, you notice that Mr. Veil has a hankering for garlic-laden cabbage laced with cayenne peppers, poured into a wooden barrel, buried like a rotting corpse, and left in the heat to ferment for weeks. No wonder he never laughs. It's impossible to eat this stinking bouillabaisse and have a sense of humor.

Mr. Veil is reading a newspaper. "Sir, there's been an incredible find at the Museum of Natural History! One of their pre-Columbian artifacts has been identified as a fossilized Pentium computer chip. Today the President announced that 'in response to this discovery, the United States is initiating a high-priority security program to limit time travel.'"

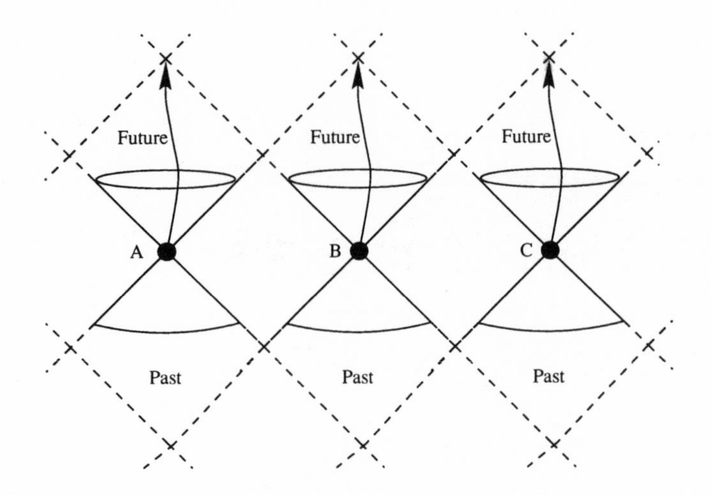

Figure 14.1 A set of three light cones for spacetime events *A*, *B*, and *C*. It is impossible to travel from any of these events to any of the others.

"We've got to be careful," Constantia says. "People are thieves. They're selling ancient artifacts. Valuable ones. The newspapers are talking of black markets in Babylonian clay tablets, Mayan and Costa Rican jade ornaments, and gold eagle pendants from the Veraguas area of Panama. They look so new that they could all be fakes. No one knows for sure if they're actually stolen from the past."

You nod, trying to ignore the stench coming from Mr. Veil's meal. "We've got to hurry with our lessons. Let's talk about backward time travel in a Gödelian universe."

Constantia nods. Today her shirt is a montage of numerous individuals with their facial colors distorted to give them purple foreheads and green hair. Most of the people were writers of popular science or science-fiction: Martin Gardner, Carl Sagan, Arthur C. Clarke, Robert Heinlein, Stephen J. Gould, and Isaac Asimov, to name a few. Constantia's stretch pants are an impossible shade of pink. She also wears jogging shoes and shocking purple socks. In other words, she makes your eyes water from thirty paces.

You bring out a funnel tied to the end of a string, and begin to whirl it around your head. "Perhaps the most interesting example of a brilliant mathematician studying cosmic questions is Kurt Gödel. He was an Austrian mathematician who lived from 1906 to 1978. Not only did he formulate a mathematical proof of the existence of God and make

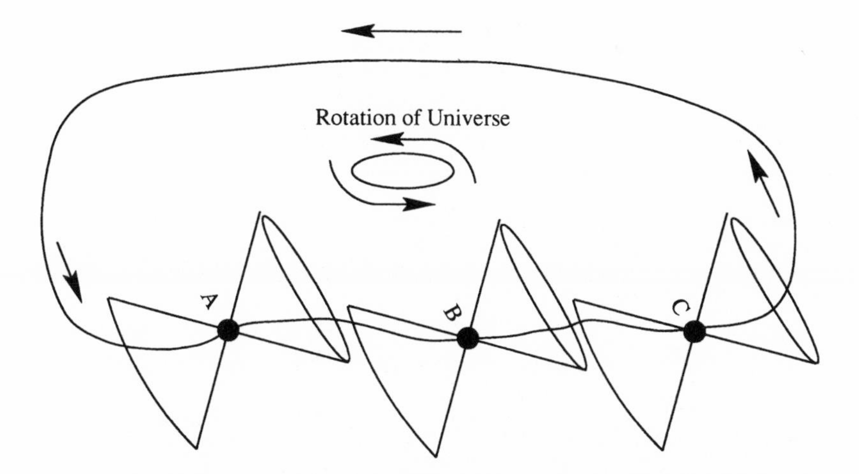

Figure 14.2 Tilted light cones in a rotating universe. Here one can travel from *A* to *B* to *C* and so on around the universe, back to event *A*. In other words, you can travel back to the same place and time that you started from without superluminal speed.

shocking contributions to pure mathematics, but he also proved that time travel is possible."

Constantia watches the funnel whirling about the table. "He proved time travel is possible?"

"Yes. He relied on Einstein for starters—Einstein's general relativity suggests that rotation of matter distorts spacetime causing light cones to tip."

Unfortunately, the string on your whirling funnel breaks, sending the funnel sailing into Mr. Veil's odorous dish. You ignore the mishap and continue your lecture. "In particular, the future half of the light cone tilts in the direction of rotation. Let's imagine a universe that rotates about a point. It turns out that the light cone tipping effect increases as you get further from that point. At a certain distance, the future half of the light cone for a given point in spacetime will tilt into the past half of similarly tilted light cones for nearby points." You sketch two diagrams (Figs. 14.1 and 14.2). "In the first diagram is a set of three light cones belonging to spacetime events *A, B,* and *C*. It's impossible to travel from any of these events to any of the others. In the second diagram, I'm drawing a family of light cones in a rotating Universe. The rotation causes a twist in spacetime that causes the cones to tip." You pause. "Mr. Veil, let's assume that you are trying to go back in time in a Gödelian universe. You can see from my drawing that the tipped cones allow you to move around a circular path on a trip into your local future and finally end up in your global past."

Mr. Veil bangs his spoon on the table. "And I can do it without ever going faster than light!"

You nod. "Correct. The roundtrip never becomes spacelike. Your path is an example of a closed timelike curve. Your path is always inside a local light cone."

A waiter comes to your table and pours some more water.

He is distinguished, about sixty years old, with a flowing gray beard and burning eyes.

You glance away from Mr. Veil's breakfast and try to concentrate on your own: fried smoked country ham, two fresh eggs over easy, grits, a buttermilk biscuit dripping with butter and slathered generously with homemade strawberry preserves. You pour a thick Vermont cane syrup on top of the biscuits. Yes, it is a cholesterol nightmare guaranteed to repulse Constantia, but you have it only once a month—so what the hell.

Constantia finishes up her bowl of cornflakes in skim milk. She traces her finger along the circle in your diagram. "Can we travel in a circle to visit Chopin?"

You shake your head. "There's a problem. The Gödelian universe in which this idea could work is infinite in size and not expanding. However, we believe that our universe is nonrotating and expanding, so while Gödel's analysis satisfies Einstein's equations, it seems unlikely that we can use it for time travel in our universe. If we wanted a Gödelian universe, it must rotate fast enough to counter the gravitational tendency to collapse. Physicists calculate that such a universe would have to rotate once every 70 billion years, and the radius at which the cones would tip sufficiently is 16 billion light-years!"

Mr. Veil finally removes your funnel from his meal and gingerly places it on the table. "Great Yggdrasill, do you mean the orbit in your drawing would be over 100 billion light-years in length?"

"Yes. To make the trip in a reasonable length of time, you'd have to be moving at near light speed. Of course, you could send radio signals traveling at light speed back in time in a Gödelian universe. That could be quite interesting. You could send a message to your grandfather."

You signal the waiter for the check. He produces one out of midair; you keypunch your credit code into it, set it for the standard gratuity, and add your thumbprint.

"More coffee?" the waiter asks. "Perhaps the lady would enjoy a liqueur? Compliments of the chef?"

You look at Constantia. "Constantia?"

"No thanks," she says.

In a few minutes, the three of you are walking down Park Avenue with its lush tree-lined sidewalks. A steady flow of people passes by: prostitutes, pickpockets, priests, pilgrims, prestidigitators, pedarasts, pimps, pill-poppers, and peddlers of all types. Park Avenue is world-renowned as the place where anything is for sale.

You walk for a few blocks as the crowd of people dissipates. Out of the corner of your eye you see replicas of you, Mr. Veil, and Constantia! Could they be future versions of yourselves who have traveled in time to this location?

Your heart is beating fast. You shake your head, and the replicas are gone. It must have been your overactive imagination. Neither Mr. Veil nor Constantia noticed anything strange.

You see a movement to your right. "What the hell is that?"

Constantia stops walking. "It could be goats."

"What? Goats?"

"Yeah," she says with satisfaction. She seems to like the fact that she is giving you information for a change. "New York has some ratty-looking herds out here that are seldom seen. The natives call them ghost herds because they seem so skittish. They have genetically engineered camouflage fur that helps them blend into the trees. The second you see them they disappear into the underbrush like smoke."

You begin to speculate. "They're not indigenous to this area. Where could they have come from?"

She laughs. "Well, the New York legend has their origin in the year 2020 when a goatherder lost his apartment and adjoining alley-way to a local bank. They say he drove his herds into the trees." She motions to the five-mile long traffic island in the center of Park Avenue that was filled with shrubs and trees. "He took his dog and shotgun and sometimes took potshots at the police, which discouraged them from trying to search for him. Some of my friends say they've seen the goatherder, just for a second, when the moon is full."

You hum a few bars of the *Twilight Zone* theme song, and you hear Constantia laugh.

"Make fun if you want, but I've seen the goats."

You glance at your watch; you have only an hour before dark.

Mr. Veil is looking closely at the bark of a tree. "Come over here," he says. "What's this?"

You see small holes leading to tunnels in the tree. With a yawn you say, "Maybe a woodpecker made them."

"Look like wormholes to me," Mr. Veil says.

You yawn again. In the distance you hear the bleating of goats. "Wormholes. That's the subject of tomorrow's lesson." You pause. "Let's go home. This is getting creepy."

The Science and History Behind the Science Fiction

> And from my pillow, looking forth by light,
> Of moon or favouring stars,
> I could behold the antechapel where the statue stood
> Of Newton with his prism and silent face,
> The marble index of a mind forever
> Voyaging through strange seas of thought, alone.
>
> —William Wordsworth, *Prelude,* Book 3, lines 58-63

Physicists and mathematicians tell us that time travel to the past should be possible under certain circumstances. For example, Kurt Gödel—one of the greatest mathematicians who ever lived — published a model of a rotating universe as a solution to Einstein's field equations for general relativity. Gödel wrote that it is "theoretically possible in these worlds to travel into the past, or otherwise influence the past." In a rotating universe, the paths (world lines) of space travelers can always move into their local future but nevertheless arrive back in their own past. You can journey along closed timelike curves (CTCs) in spacetime. In Gödel's classic 1949 essay, he was aware that his rotating Universe model created paradoxes:

> By making a round trip on a rocket ship in a sufficiently wide course, it is possible in these worlds to travel into any region of the past, present, and future, and back again, exactly as it is possible in other worlds to travel to distant parts of space. This state of affairs seems to imply an absurdity. For it enables one, for example, to travel into the near past of those places where he has himself lived. There he would find a person who would be himself at some earlier period of his life. Now he could do something to this person which, by his memory, he knows has not happened to him.

In short, Kurt Gödel found a disturbing solution to Einstein's equations that allows time travel.[1] For the first time in history, backward time travel had been given a mathematical foundation! Isaac Newton thought of time as a chariot traveling along a straight highway. Nothing could deflect or change the course of the vehicle once it had started. He believed in one all-embracing universal time. Time could not be affected by anything; it just went on flowing at a uniform rate. Einstein, however, showed that the road on which the chariot traveled could curve, although it could never loop back on itself like a circular race track. He showed that the Newtonian concept of universal time yields absurd conclusions when applied to the behavior of light signals and the motion of material objects. Gödel went even further and showed that the road could circle back on itself. The main problem with Gödel's model is that it requires the universe to be slowly rotating, and astronomers see no evidence of this. However, *if* our universe indeed rotated, then CTCs and time travel would be physically possible. That's quite a mind-boggling finding. Even though impractical, general relativity predicts the possibility of time travel where creatures could visit or signal their grandparents by traveling around large circular orbits.

In the Gödelian universe, the tendency for collapse due to gravitational self-attraction is exactly balanced by the centrifugal force due to universal rotation. Gödel's model must make one complete turn every 70 billion years to function as a natural time machine. (Our universe appears to be 10 to 20 billion years old.) If the universe were spinning at the Gödel rate, our universe's critical radius would be about 16 billion light-years, and the shortest CTC would be about 100 billion light-years. This means that you would have to travel this distance in order to return to the present. Such a trip could be made in one subjective year or less if your ship traveled sufficiently close to the speed of light that the length of the CTC was appropriately

Lorentz-contracted. This speed would certainly require a huge amount of energy, but perhaps electronic signals could circle a Gödelian CTC at a cheaper cost.

Would it be possible to send a signal to our past? As I just suggested, our universe is likely not Gödelian because it appears to be finite, expanding, and nonrotating. Gödel's universe is infinite, static, and rotating. Even if our universe were Gödelian, it is so small there is not enough room to make the minimum 100 billion light-year CTC travel loop. If we wanted our small universe to create a Gödelian CTC, it would have to rotate more than ten times faster than the minimum Gödel rate.

Einstein was quite aware of the awesome nature of Gödel's findings. In 1949, Einstein wrote:

> Kurt Gödel's essay constitutes, in my opinion, an important contribution to the general theory of relativity, especially to the analysis of the concept of time. The problem disturbed me at the time of the buildup of the general theory of relativity, without my having succeeded in clarifying it. . . . [T]he distinction "earlier–later" is abandoned for world-points which lie far apart in a cosmological sense, and those paradoxes, regarding the *direction* of the causal connection, arise, of which Mr. Gödel has spoken. . . . It will be interesting to determine whether these are not to be excluded on physical grounds.

Over the years, other time-travel solutions have been discovered, in the framework of general relativity, that violate causality. In other words, time travel was found not to be an anomaly of Gödel's particular analysis but rather was built into the basic gravitational field equations of general relativity.

Kurt Gödel pondered other cosmic questions in addition to time travel. Sometime in 1970, Gödel's mathematical proof of the existence of God began to circulate among his colleagues. The proof was less than a page long and caused quite a stir, and I discuss the proof in my book *The Loom of God.* Despite his unusual excursions into theology, Gödel's academic credits were impressive. For example, he was a respected mathematician and a member of the faculty of the University of Vienna starting in 1930. He also was a member of the Institute of Advanced Study in Princeton, New Jersey. He emigrated to the United States in 1940.

Aside from time travel and God, Gödel is most famous for his theorem that demonstrated there must be true formulas in mathematics and logic that are neither provable nor disprovable, thus making mathematics essentially incomplete. (This theorem was first published in 1931 in *Monatshefte für Mathematik und Physik*, volume 38.) Gödel's theorem had quite a sobering effect upon logicians and philosophers because it implies that within any rigidly logical mathematical system there are propositions or questions that cannot be proved or disproved on the basis of axioms within that system, and therefore it is uncertain that the basic axioms of arithmetic will not give rise to contradictions. The repercussions of this fact continue to be felt and debated. Moreover, Gödel's article in 1931 put an end to a century's long attempt to establish axioms that would provide a rigorous basis for all of mathematics.

Over the span of his life, Gödel kept voluminous notes on his mathematical ideas. Some of his work is so complex that mathematicians believe many decades will be required to decipher all of it. Author Hao Wang writes on this very subject in his book *Reflections on Kurt Gödel* (1987):

> The impact of Gödel's scientific ideas and philosophical speculations has been increasing, and the value of their potential implications may continue to increase. It may take *hundreds of years* for the appearance of more definite confirmations or refutations of some of his larger conjectures.

Gödel himself spoke of the need for a physical organ in our bodies to handle abstract theories. He also suggested that philosophy will evolve into an exact theory "within the next hundred years or even sooner." He even believed that humans will eventually disprove propositions such as "there is no mind separate from matter."

Gödel was fascinated with time travel due, in part, to his failing health. He was terrified at the thought of death. We can speculate that his interest in a Gödelian universe sprung from his hopes of avoiding death and reliving his life.

Among all the great discoveries of the last five hundred years, to me, at any rate, the biggest, most marvelous discovery of all is the discovery of how life came into being—the discovery which we associate with the name of Darwin and DNA. Two hundred years ago you could ask anybody, "Can we someday understand how life came into being?" and he would have told you, "Preposterous! Impossible!" I feel the same way about the question, "Will we ever understand how the universe came into being?" And I can well believe that the evidence that we need is right in front of us, right now. We just have to look in front of our noses.

—John Archibald Wheeler

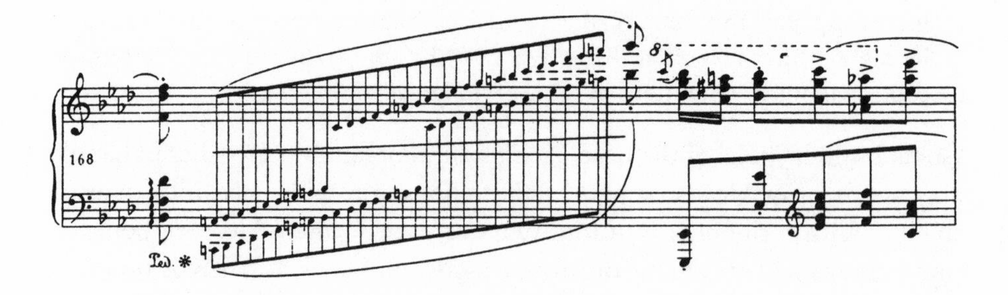

Luck is an essential part of a career in physics.

—Leon Lederman

We do not really change over time; we are as flowers unfolding; we merely become more nearly like ourselves.

—Anne Rice, *The Queen of the Damned*

Wormhole Time Machine

15

✍ At dinnertime, Trump Tower restaurant on Fifth Avenue is packed with local artists talking shop over plates of steaming squid or—Trump's specialty—bacalao in a thick black bean sauce. Despite the latest health craze, some of the artists are really enjoying their meals, taking off their fringed vests and rolling up the sleeves on their Nehru jackets, enjoying the oily food. An old-fashioned wood and wicker fan spins over your heads. It makes a pleasant change from the arctic air conditioning in some parts of the Museum of Music.

Outside, the mood is less congenial. There are screaming protestors, and Zetamorphs with megaphones. "Ban FTL," their signs read. One girl with flowers on her shirt and headband holds a huge sign with the blood-red words, "Down with the Reinterpretation Principle." A little further down the street you see a vendor selling fragments of wood purported to be from the cross on which Jesus was crucified. It has to be a hoax. However, should time travel ever become common, the black market trade in historical artifacts would be overwhelming, and Jesus' cross would be reduced to mere splinters. In your mind you see an image of human piranhas circling all the famous ancient artifacts—the Cross, fragments of JFK's brain, Hitler's moustache, Moses' staff, the horn of Abraham's ram, the bones of martyrs . . .

Mr. Veil turns to you. "Sir, if time travel is possible, it seems that every time you go back and talk to your former self, and change your future actions, you'd be creating a duplicate of yourself in an alternate world. However, the person wouldn't quite be the same as you because his memories would be different from yours. Would the two of you be the same person?"

"It seems logical that you could duplicate slightly different versions of yourself over and over again." You pause. "Maybe we can do it with wormholes."

"Wormholes?" Constantia says.

You nod. "Wormholes might be used for time machines. Physicists believe that at the heart of all space, at submicroscopic size scales, there exists quantum foam. If you were to magnify space it would become a seething, probabilistic froth—a cosmological milkshake of sorts."

Constantia runs her fingers lingeringly through her hair. "Now, this sounds quite interesting. What do we know about the quantum froth?"

Your heartbeat increases in frequency and amplitude as you gaze at her skirt the color of cool mint, with tiny ruffles at the breast, leaving her shoulders bare except for thin spaghetti straps. "In the froth," you say, "space doesn't have a definite structure. It has various probabilities for different shapes and curvatures. It might have a 50 percent chance of being in one shape, a 10 percent chance of being in another, and a 40 percent chance of being in a third form. Because any structure is possible inside the froth, we can call it a *probabilistic foam,* or *quantum foam.*"

You whip out a pocket computer from your back pocket and press the enter key. A probabilistic froth undulates on the screen (Fig. 15.1; Code 4 in Appendix 2). You look deeply into Constantia's eyes, which appear mesmerized by the undulation. "This computer model of two-dimensional froth produces tunnels and wormlike tubes commonly depicted in schematic diagrams for quantum froth. The connecting bridges in the foam correspond to *wormholes* between different universes or between different places and times in the same universe."

Figure 15.1 Two-dimensional computer model of a quantum foam. The algorithm in Code 4 in Appendix 2 produces seething probabilistic froth and tunnels. Quantum foam is thought to exist at submicroscopic scales where the geometry and topology of space are probabilistic.

Constantia's mouth hangs open. "How did you make such beautiful shapes?"

"The simulation is performed using a checkerboard world. Secret algorithm from the 20th Century."[1]

She pouts. "Please, can you tell me?"

You smile. "It's based on a cellular automaton consisting of a grid of cells that can exist in two states, occupied or unoccupied. The occupancy of one cell is determined from a simple mathematical analysis of the occupancy of neighbor cells. Though the rules are simple, the patterns are very complicated and sometimes seem almost random, like a turbulent fluid flow or the output of a cryptographic system."

You signal for the waitress. She is starvation-slim with painfully sharp cheekbones and a scraggly mop of tangled maroon hair. You and Constantia order large seafood salads. Constantia avoids the bread—watching her waistline, you suppose. Veil, unfortunately, orders some strange Zetamorph dish loaded with ethyl mercaptan and butyl selenomercaptan, the most evil-smelling substances known to man.

They remind you of a combination of rotting cabbage, garlic, onions, burnt toast, and sewer gas.

Constantia begins to choke.

"Sorry," Mr. Veil says as he pulls his meal as far as possible from you and Constantia, probably hoping it will diminish the stench.

You briefly pace around the table. "Quantum foam is everywhere, in the tiny centers of black holes, in space, even in your body. But you'd need a powerful microscope to examine it. We're talking size scales around 10^{-33} centimeters."

Mr. Veil's forearms begin to quiver as he looks at you. "Can we—?"

"Mr. Veil, I've always wanted to enter a black hole or a wormhole and travel to another universe."

Mr. Veil's stiff forelimb taps repeatedly on the floor. "Sir, this reminds me of my old friend, Mr. Plex—the famous black hole researcher. He and I have talked about black holes in the past. They might contain holes to other regions in space and time."

You nod and sit down. "That's right, Mr. Veil. Such a wormhole could also be connecting different areas of our same universe." You pause. "I could survive if the black hole were large enough. The tidal force is actually less for larger holes. Right at the edge a black hole with mass M, the relative acceleration between my head and foot would be proportional to $1/M$."

"But sir, the tidal gravity will still crush you as you get near the black hole's throat." Mr. Veil pauses. "I also hear rumors that the hole between universes is very unstable. Once it forms, it expands and contracts before anything can cross it."

You stop chewing for a second and then fork a big fat shrimp. "That's why I've been considering diving into a *rotating* black hole. Instead of the *point singularity* (the dense microscopic center of a static black hole), a rotating black hole has a *ring-shaped singularity,* a cosmological doughnut of sorts. I could go through the ring without encountering infinitely curved spacetime!" You pause. "After going through the ring, I'd emerge into another region of spacetime, usually interpreted as another universe. It's even possible to adjust my trajectory through the tunnel to emerge whenever I wish, even thousands of years from now."

Constantia gazes into your eyes. "I sense a hesitation in your voice."

"Constantia, some scientists believe that the ring singularity spews an intense flux of high-energy particles into the tunnel between universes. Not only might they kill me, but they might seal off the tunnel."

Constantia comes a bit closer and puts her hands on her hips. "Then what do you propose?"

"Think about quantum foam. In the foam, adjacent regions of space are continually stealing and giving back energy from one to another. These cause fluctuations in the curvature of space creating microscopic wormholes."

The tapping of Mr. Veil's forelimb becomes more incessant. "But sir, those wormholes are too tiny for you to use."

"That's the problem, and why we won't try to use a wormhole. We'd need a device that spews out something called *exotic matter*. Exotic matter will enlarge and hold open a wormhole. Maybe some advanced extraterrestrial civilization has such a device, but we don't know. We may never get to use a wormhole for time travel. But certainly we can do some computer experiments. A traversable wormhole can be described by

$$z(r) = \pm\, b_0 \ln\left(r/b_0 + \sqrt{(r/b_0)^2 - 1}\,\right)$$

The schematic diagram has two surfaces and two asymptotically flat regions far from the hole. The throat is smallest at a minimum value of r, where $r = b_0$. The two flat regions correspond to two universes."

Constantia withdraws a notebook computer from beneath a potted philodendron plant near your table. She starts to hand it to Mr. Veil, but hesitates. "Let me try programming it," she says as she begins to type (Code 5, Appendix 2). After a few seconds a double trumpet shape emerges on the screen (Fig. 15.2).

A shiver runs from the very base of your spine up to your neck. You've never seen her type so fast. "Notice that the radial coordinate r has a particular significance. If you place a circle around the wormhole's throat, $2\pi r$ is the circumference. r decreases from $+\infty$ to a minimum value of b_0 as one moves through one universe to the other."

Constantia takes your hand in hers. Her eyes are bright. "What would the worm hole really look like?"

You give her hand a squeeze. "The diagram is just a convenient way to represent four dimensions on a computer screen. In reality, both mouths of the worm hole would look like two spheres floating in our

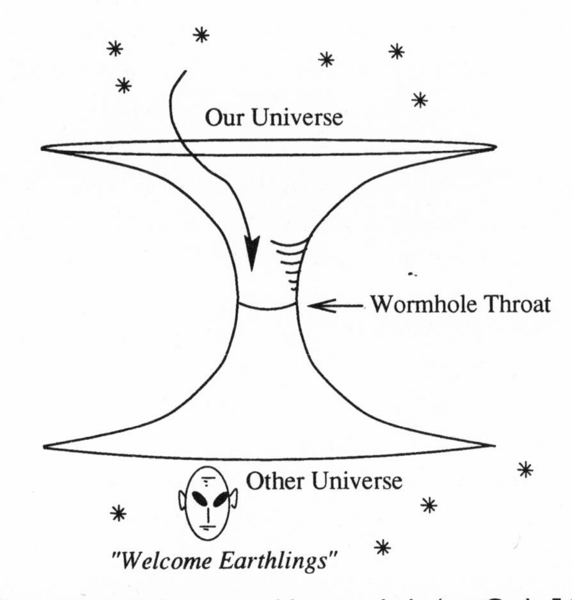

Figure 15.2 Cross-section of a traversable wormhole (see Code 5 in Appendix 2). Here the wormhole connects two disjoint universes, as opposed to the wormhole in Figs. 15.3 and 15.4, where the wormhole is a connection in the same universe. For computer hobbyists among you, the formula used to construct the wormhole is $z(r) = \pm\, b_0 \ln(r/b_0 + \sqrt{(r/b_0)^2 - 1})$.

three-dimensional universe. Of course, we might not be able to see their spherical shapes at all."

"But how could you use this as a time machine?"

"One way is to move one mouth of the wormhole relative to the other. Maybe we can use a large asteroid's gravitational attraction to drag on the end of the wormhole at high speed and create the time dilation effect." You pause. "Remember, we spoke of the fact that a moving clock runs slow with respect to a stationary clock." You begin to sketch on the tablecloth (Fig. 15.3). You look back at Constantia. "Here's a diagram of a wormhole. Suppose we place clock 1 at one mouth of the wormhole and clock 2 at the other mouth. Both report the same time, as do nearby clocks. Next, suppose we place mouth 2 on a rocket ship that takes a long, high-speed trip out into space and returns. We can keep the wormhole handle short even after a long trip just as we could keep two points on a piece of paper close if we fold the paper." You pause and draw another sketch showing our universe represented as a curved piece of paper (Fig. 15.4). Then you return to the

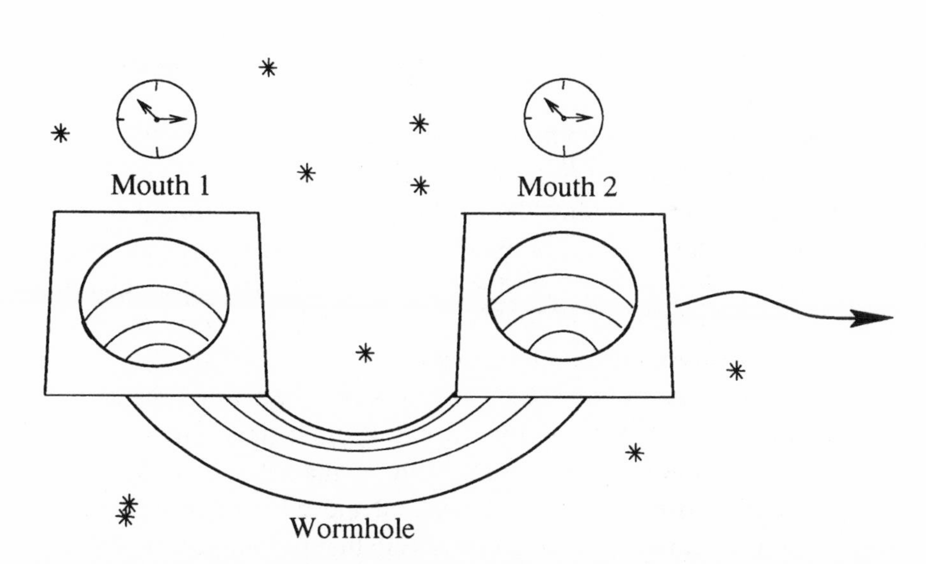

Figure 15.3 Building a wormhole time machine. Like the other wormhole figures in this chapter, this diagram is schematic because it is a two-dimensional rendition of a wormhole connecting two places in three-dimensional space. Time machine wormholes actually connect two locations in four-dimensional spacetime, and the mouths would appear to be three-dimensional spheres rather than depressions in a plane.

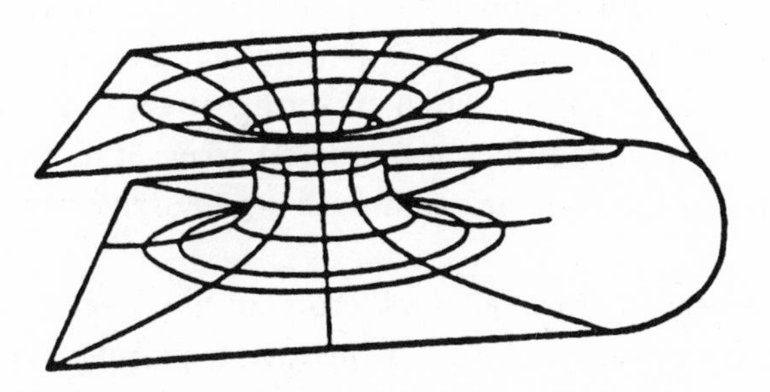

Figure 15.4 Schematic diagram of a wormhole connecting two regions in the same universe represented by a curved piece of paper with depressions in it. Notice that distance between the mouths of the wormholes can be made far apart in our universe (by moving the holes to the left along their respective planes), although the hyperspace wormhole connection can be quite short.

previous diagram. "After the ship has returned we unload mouth 2 from the rocket ship."

You are distracted for a moment as your gaze wanders over various sculptures in Trump's gallery windows. Quite impressive. The oldest work appears to be a ceramic female figurine from the coastal Ecuadorian Valdivia culture (ca. 2000 B.C.). Two fine Tairona blackware effigy urns from Colombia highlight the collection. Peruvian Chavyn blackware, burnished pieces of the Nazca, and Moche ceramics create an engaging view of pre-Conquest America.

Constantia follows your gaze. "Trump always had taste." She pauses for a moment and returns to the wormhole discussion. "Since the distance between mouth 1 and mouth 2 can be small, the clocks can actually stay synchronized, but since clock 2 moved with respect to its external space, it reads earlier than clocks in the space outside mouth 2."

Mr. Veil says, "This is like the twin paradox we discussed. Each mouth plays the role of one of the twins."

"Right, now suppose that the journey of mouth 2 produced a two-hour time difference between clock 2 and its local, external clocks. This means that if clock 2 reads 10 a.m., the clocks outside of mouth 2 will read 12 noon. But remember that clock 2 and clock 1 read the same time, as do the clocks in the vicinity of clock 1. Thus, the wormhole connecting 2 and 1 connects two parts of our universe which are two hours apart." You pause. "Now assume that a trip from mouth 1 to mouth 2 takes one hour when traveling outside the wormhole. Then you could leave mouth 1 at noon and arrive at mouth 2 at 1:00, enter mouth 2 and travel back to mouth 1 and arrive before the trip began!"

"Great Yggdrasill!" Mr. Veil says. "You could repeat the same trip and continue to go back an hour during each trip! But you couldn't go back in time before the wormhole time machine's creation."

"That's correct. But enough about wormholes for now. I don't have a way to produce the necessary exotic matter to create a large wormhole. Even if I could create a wormhole today, I couldn't use it to travel to a time prior to today."

Constantia is deep in thought. "I have an idea," she says. I want to take my camera back in time when we visit Chopin and get some good pictures. Then I can write up our trip as a feature."

"A feature? Feature for what?"

"To sell, silly. It's exactly the kind of material they're looking for in the big travel magazines. Maybe *Time* magazine would like it."

Mr. Veil shakes his head. "Better publish the photos under a pseudonym or the Time Cops might throw you in jail."

You consider Constantia's suggestion and look at her. "Well," you pause. "I was thinking this might be a private trip for both of us—not a working trip. We might not want to call too much attention to ourselves."

Constantia nods. "Why don't we wait and see?"

"Okay." You slap your hand at a robotic mosquito that has come under the eaves of Trump Tower to get out of the rain. Like many computer viruses, the robotic mosquitos were engineered by teenaged rogues trying to show the world their intellectual prowess while making nuisances of themselves. The mosquitos usually don't bite you, but their oscillating whine is more than you can take when combined with the stench of Mr. Veil's food.

You try to ignore the mosquitos and look out the window. It has been raining since the afternoon, rain that falls in a light curtain over Fifth Avenue, but the rain brings little freshness or relief. You stare into Constantia's eyes, one hand raised against the screaming mosquitos, the other holding your nostrils closed. Mr. Veil's food is making you sick.

"I'm finished," says Mr. Veil. From beneath the table he whips out a purple-handled woman's umbrella and fusses with the clasp. "Shall we go?"

Constantia signals the waiter to take Mr. Veil's dish of mercaptans while Mr. Veil continues to fumble with the umbrella, occasionally banging it into your chest.

"Give me that thing," you scream at him.

Suddenly a motion on the table catches your eye. "The salt is moving," you whisper.

Constantia looks closer. "No, wait, I don't think so. They're ants."

You roll your eyes. "What next? I can't take much more of this day."

Mr. Veil whips a magnifying glass out of his hip pocket and hands it to you. "I'm familiar with these ants."

You take the magnifying glass and look closer. Except for their heads, which resemble hornets' heads, the ants seem to be nothing more than silicon-like chips with moving legs. On their backs are solar cells that control their microprocessor brains. You watch as a six-legged insect walks as a terrestrial ant does, always lifting the middle leg on one side together with the front and back legs of the other.

"What are they made of?" you ask Mr. Veil.

"The legs are composed of zinc oxide," he says brusquely, as if he is insulted by your grabbing his umbrella. "It's a piezoelectric material that expands and contracts when exposed to a voltage. These mechanical legs extend and curl in response to tiny electrical currents supplied by the microprocessor brains."

"How do you know so much about this?"

"Sir, I placed them there. Thought you would like —"

"Mr. Veil! Constantia and I were trying to have a nice dinner."

Mr Veil's own piezoelectric mouth grins as he pops a few ants into his mouth.

"My God," Constantia screams. You hear a crunching sound, as the zinc-oxide ants are devoured alive. The ants make a horrible whining sound. A few legs dangle from Mr. Veil's elliptical mouth.

You quickly throw down some money on the table, grab Constantia's hand, and head for the door. "C'mon, let's get out of this nuthouse." ✍

The Science Behind the Science Fiction

> If we [discoverers of the DNA double helix] deserve any credit at all, it is for persistence and the willingness to discard ideas when they became untenable. One reviewer thought that we couldn't have been very clever because we went on so many false trails, but that is the way discoveries are usually made. Most attempts fail not because of the lack of brains but because the investigator gets stuck in a cul-de-sac or gives up too soon.
>
> —Francis Crick

As we discussed in the last chapter, Gödel's time machine, proposed in 1949, worked on huge size scales—the entire universe had to rotate. At the other extreme are cosmic wormholes created from subatomic quantum foam as proposed by Kip Thorne and colleagues in 1988. Not only did these researchers claim that time travel is possible in their prestigious *Physical Review Letters* article, but time travel is probable under certain conditions.

In their paper, they describe a wormhole connecting two regions that exist in different time periods. Thus, the wormhole may connect the past to the present. Since travel through the wormhole is nearly instantaneous, one could use the wormhole for backward time travel. Unlike the time machine in H. G. Wells's *The Time Machine* the Thorne machine requires vast amounts of energy to use—energy that our civilization cannot possibly produce for many years to come. Nevertheless, Thorne optimistically writes in his paper: "From a single wormhole an arbitrarily advanced civilization can construct a machine for backward time travel."

Note that the term "wormhole" is used in two different senses in the physics literature. The first kind of wormhole is made of quantum foam. One result of the foamlike structure of space is the likely existence of countless wormholes connecting different parts of space, like little tubes. In fact, the theory of "superspace" suggests that tiny quantum wormholes must connect every part of space to every other part! The other use of the word "wormhole" refers to a possible zone of transition at the center of a rotating black hole.

Various debates continue with respect to theoretical possibility of the wormhole time machine. For readers interested in other discussions, see Visser's papers.

The quantum foam diagram in Figure 15.1 is discussed in my book *Black Holes: A Traveler's Guide* (Wiley) where I also include color graphics of three-dimensional versions of the undulating froth. For additional information, see note 1.

We cannot yet answer the ultimate questions, but we can discuss the questions intelligently.

—Alan Guth

How wonderful that we have met with paradox. Now we have some hope of making progress.

—Niels Bohr

It is impossible to travel faster than the speed of light, and certainly not desirable, as one's hat keeps blowing off!

—Woody Allen

258 *ff*

Man is not born to solve the problems of the universe, but to find out where the problems begin, and then to take his stand within the limits of the intelligible.

—Goethe

The dog hears some sound from the farmer's house and thinks of the shotgun with its wide black holes like a figure eight rolled onto its side. The dog knows nothing of figure eights, but even a dog may recognize the dim shape of eternity if its instincts are honed sharp enough.

—Stephen King, *Four Past Midnight*

Adventures with Time

16

✍ "Wait!" Constantia screams as you plunge a needle into her upper arm. "Are you sure this is necessary?"

You nod. "Preparation for our time travel. Inoculation against germs."

"What?" she says, rubbing her arm.

"When you travel from one time to another you inevitably expose the inhabitants to new microbiological environments to which their immune systems are not adapted. When Europeans and Native Americans met, epidemics often occurred. A virus or bacterium that gave a Mayan a cough could kill a Spaniard, and vice versa."

You walk with Constantia and Mr. Veil through a large room, down a pegged-pine corridor covered with Persian rugs, and through narrow hallways lined with framed eighteenth-century prints of famous composers. After another thirty feet, you enter a massive atrium with a series of marble pedestals on which stand busts of Mozart, Beethoven, Bach, Brahms, and Eck—an alien blob-like mass with dozens of throat appendages used for playing musical instruments.

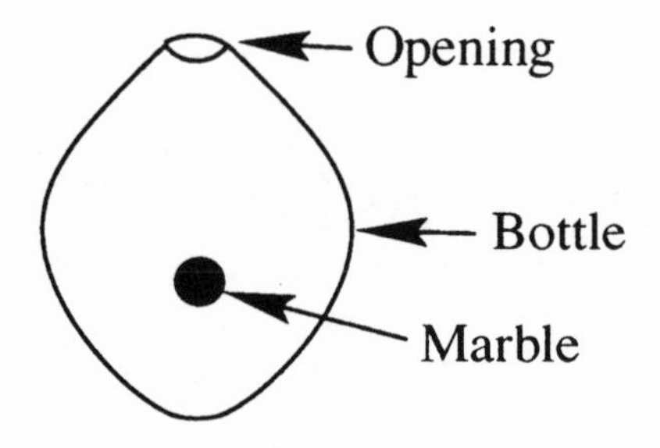

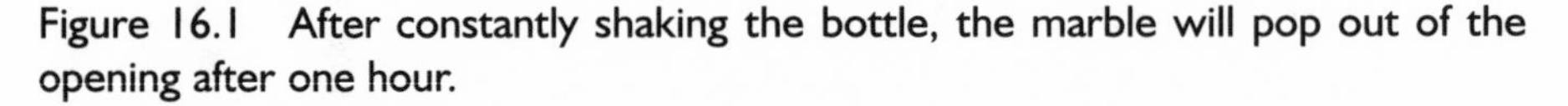
Figure 16.1 After constantly shaking the bottle, the marble will pop out of the opening after one hour.

Constantia looks at a paper that has come off a fax machine sitting beside the bust of Beethoven. "What's this?" she asks.

On the paper is the following enigmatic transmission:

> The Maya who lived in the southern Yucatan Peninsula around A.D. 500 were an unusually gifted and cultured people. They were skilled in agriculture, architecture and art—so why did they die out? zilfmw 900 z.w., dsvm gsv nzbz ivzxsvw gsvri kvzp, gsvri xrerorazgrlm hfwwvmob xloozkhvw. hnzoo tivb yvrmth erhrgvw vzigs zg gsrh grnv, zmw girttvivw kvzhzmg ivelogh ztzrmhg gsv iformt xozhhvh. zg gsv hznv grnv, z hveviv wizftsg wvhgilbvw gsv nzbzm ztirxfogfiv. hxrvmgrhgh vcznrmrmt hvwrnvmgh zg gsv ylggln lu ozpv xsrxszmxzmzy rm xvmgizo bfxzgzm szev ulfmw zorvm zigruzxgh yvgdvvm 800 zmw 1000 z.w. zugvi vcznrmrmt gsv izgrl lu lcbtvm zs gl liwrmzib lcbtvm zu rm znksrlcfh zmw hmzro hsvooh rm gsv ozpv, dv yvorvev gszg zorvmh szev xzfhvw nfgzgrlmh rm gsv olxzo droworuv—nzprmt gsv nzbz nliv hfhxvkgryov gl wrhvzhv.

You look at the paper. "No one knows what it says. Some of my agents found it in the hand of a dead Time Cop in Roswell, New Mexico. So far no one has been able to decode it. I try to translate it in my spare time."

You place the fax on a table. "I'd like to have some fun with time today. Recreations. I want to help you get a better appreciation for probability and vast amounts of time." You hand Mr. Veil a large glass bottle with a small opening at its top (Fig. 16.1). You also hand him a small blue marble. "Mr. Veil, please place the marble in the bottle and shake it randomly to see if you can pop the marble out of the upper opening."

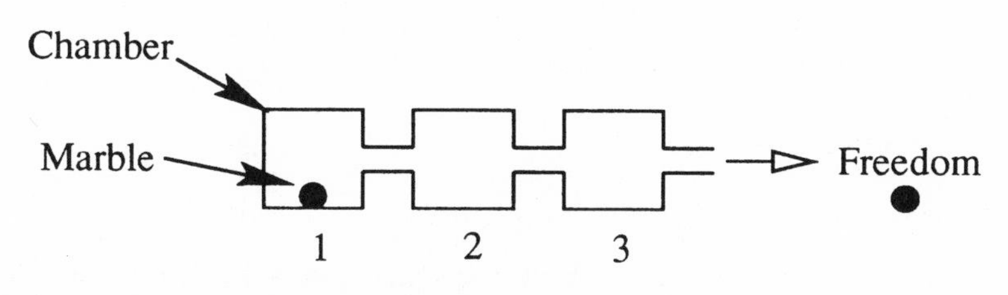

Figure 16.2 The marble will exit after six hours.

Even after violently shaking the bottle for five minutes, the marble does not leave the bottle through the tiny aperture. "Sir, my arm is tiring."

"Keep shaking. We want to find out how long it takes for the marble to pop out." You wait another five minutes.

"Sir, I can't take much more of this."

"Okay. Let's rig a machine to shake the bottle." You place the bottle into a shaking machine, and after about one hour of constant bouncing around in the bottle, the marble pops out!

You turn to Mr. Veil and Constantia. "I happen to have a legolike construction set that will allow us to do some experiments. We can construct chambers with characteristics identical to the bottle, so that on average it takes the marble one hour to pop out an opening. What would happen if I connected multiple chambers in series using their small openings? Chamber 1 would have just one opening, just as my bottle did. Chamber 2, however, would have two openings, one which connected it to chamber 1, and the other which connected it to chamber 3, and so on." You link three chambers together (Fig. 16.2).

"Let's assume that it takes one hour for the marble to find an opening, as it did in our single-bottle experiment. How long would it take for the marble, starting in chamber 1, to leave chamber 2 and jump into chamber 3? How long would it take for the marble to find its way to chamber 4? How long would it take for the marble to exit chamber *n*? Remember that in each of the intermediate bottles, the marble has just as likely a chance of moving into a previous bottle as it does moving forward. Assume this is an ideal system. It has no friction, gravity, and so on."

"Sir, I have no idea how long it would take for the marble to get out."

"It turns out that the average time to get from Chamber 1 to the *n*th bottle's opening to the outside world is approximately $n(n+1)/2$ times

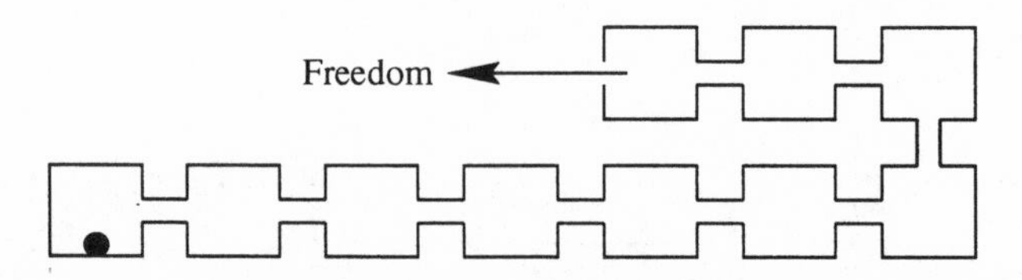

Figure 16.3 The marble will take 55 hours to leave the tenth chamber in this set.

the average time to find an opening.[1] (For a large number of bottles, you can approximate this equation by n^2.) For example, let's rig together ten chambers." You pause while you build the assembly of chambers (Fig. 16.3). "It will take 55 hours for the marble to leave the 10th bottle in this set."

You hand the assembly to Constantia. "How would you picture 5000 hours in a diagram of connected chambers?" You pause to heighten the suspense. "It turns out that 5000 hours is represented by 100 chambers. This means that it will take you a good part of a year to shake the marble out of a connected series of 100 chambers!

"What happens if we now provide *two* openings between each chamber? One opening is free-flowing, and the other has a one-way valve allowing the marble only to travel in a *backward direction*—a direction away from the opening of final egress. The average time for the marble to get out of the *n*th bottle in the case of one forward opening and M backward openings can be approximated[2] by $2 \times (M+1)^n / M^2$."

Constantia's eyes wander as she contemplates the new arrangement. "This should really slow down the marble's final escape."

"Yes, the addition of multiple one-way backward valves between the connected chambers increases the time for the marble to exit because the marble has a better chance of going backward than forward. To facilitate our discussion of the incredible characteristics of these connected chambers, I've invented some nomenclature. Rather than say, 'three chambers with two backward openings and one forward opening between them,' I'll simply use the symbol $C(3,2)$. In all of the following discussions and examples there is always one forward connector and M backward connectors between the chambers. For example, here is a picture of a $C(9,2)$ system. Each connector is represented by a line." You sketch on a pad (Fig. 16.4).

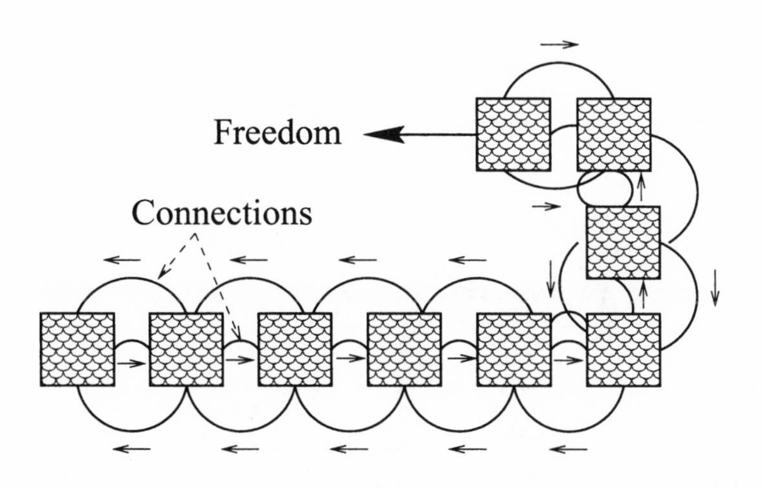

Figure 16.4 A little over a year is represented by an assembly of nine chambers connected by one forward connector and two backward connectors—because the marble takes over a year to escape. This is called a $C(9,2)$ assembly. In the following figures, there is always exactly one forward connector and several backward connectors.

"Can you guess, or calculate, how long it will take a marble to leave the last bottle of $C(9,2)$? It turns out that it will take over a year of shaking to get the marble out of the bottle chain.

"Here are some other 'time-in-a-chamber' diagrams for you to ponder. The following $C(9,5)$ collection represents the life of a human being." You sketch nine chambers each with five backward connectors and one forward connector (Fig. 16.5). "$C(15,5)$ represents roughly 1.6 million years, or the number of years ago that *Homo erectus* was born." You sketch the $C(15,5)$ system on the pad (Fig. 16.6). "*Homo erectus* (upright man) is thought to be the direct ancestor of humans (*Homo sapiens*), and a *Homo erectus* skeleton was discovered in Kenya in 1985 and dated to 1.6 million years ago. Thus, we may call $C(15,5)$ the '*Homo erectus* system.' If this ancient human began shaking the fifteen bottles and handed this arduous task down from generation to generation, only today would the marble finally pop out of the last chamber." You pause to think of more examples. "The earliest and most primitive dinosaur is the *Herrerasaurus* discovered in 1989 at the foot of the Andes in Argentina. It is believed to be about 230 million years old. A $C(15,7)$ assemblage represent this span of time. A $C(15,10)$ assem-

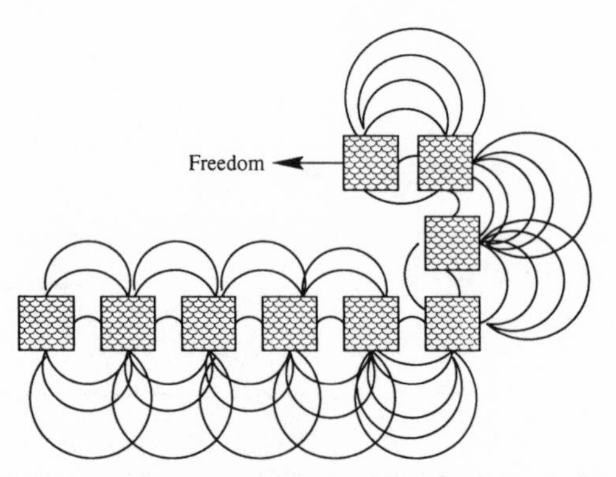

Figure 16.5 A *C*(9,5) system represents the life of a human being.

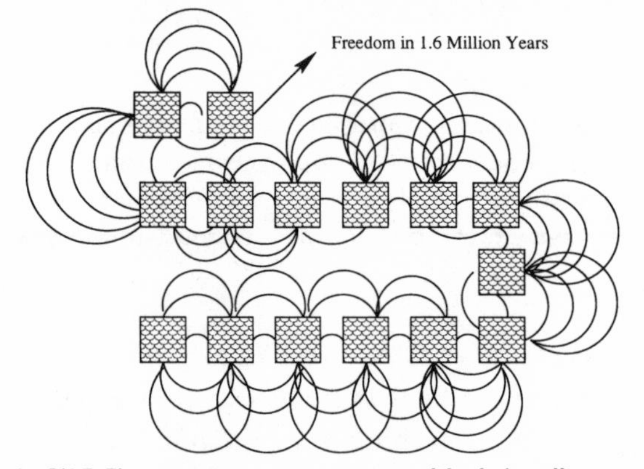

Figure 16.6 A *C*(15,5) system represents roughly 1.6 million years, or the number of years ago that *Homo erectus* was born.

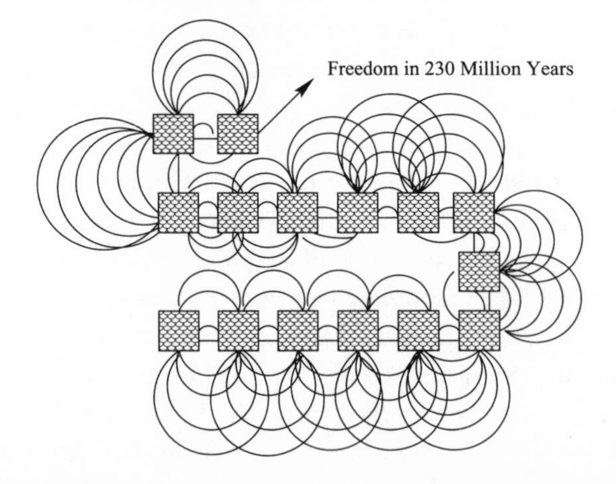

Figure 16.7 A *C*(15,7) system is a *Herrerasaurus* system, named after the earliest and most primitive known dinosaur that is believed to be about 230 million years old.

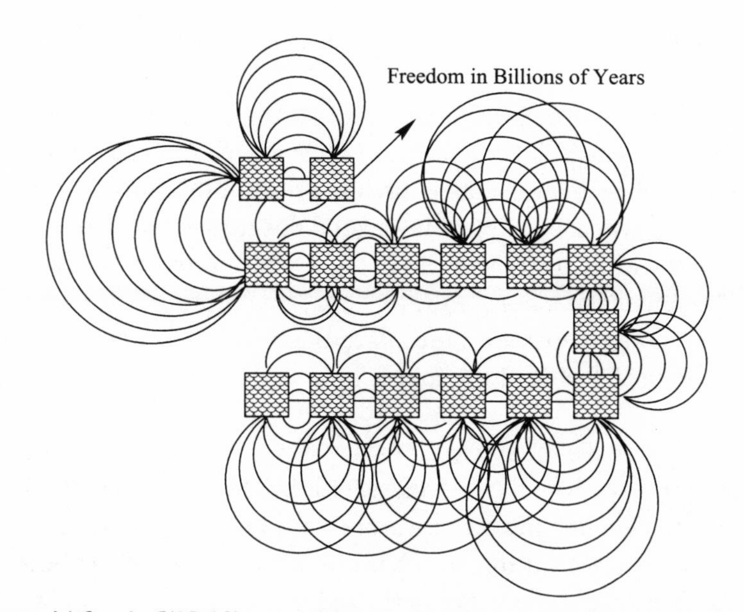

Figure 16.8 A *C*(15,10) assemblage represents more than twice the age of the Earth (4.5 billion years).

blage represents more than twice the age of the Earth (4.5 billion years)." You sketch the systems on the pad (Figs. 16.7 and 16.8).

"Incredible," says Constantia as she stretches her legs.

You start putting away the assemblies of chambers. "I think we need a break," you say.

Constantia nods as she brushes a mosquito away from her face. "How about we go to the Russian Tea Room?"

"Excellent," you say.

After about a twenty-minute walk you finally arrive at the restaurant. Constantia walks a pace behind you, as the Russian Tea Room's owner, Madame Dudevant, greets you with a kiss. The three of you are immediately whisked to the center table in the front room, where several waiters and busboys bow as they bring bread, butter, Perrier water, and a huge bowl of borsht. Constantia obviously enjoys every minute of it. So do you.

Constantia motions to the walls. "They've really spared no expense. I recognize paintings from the Pskov school of late Medieval art. They were famous in the Russian city of Pskov. Pskov and Novgorod both remained free of Mongol rule in the two centuries following the invasions of Russia in the mid-thirteenth century. That preserved and transformed the Byzantine artistic tradition that was the basis of Russian art."

You gaze at the artwork with their large color masses dominated by fiery orange-red and a deep "Pskovian" olive green. You can imagine these paintings illuminating dark church interiors of this northern city. The restaurant owners must be doing a good business to afford these.

You turn toward Constantia and bring a small sphere out of your pocket. The sphere has a hole in the top and resembles a goldfish bowl. "Constantia, take ten pennies and place them in this sphere. Try to keep them on the left side of the sphere."

"Okay."

"Now start to shake the sphere."

Covering the opening to the sphere with her hand, she shakes the bowl so that the ten pennies bounce around the bowl in random directions. As they bounce off one another and off the sides of the bowl, some leave the left side and move to the right.

"In the course of time, the distribution of pennies in the two halves will inevitably become more uniform—that is, the entropy of the system will increase. After a few seconds, the bowl will approach a state of maximum entropy, with nearly equal numbers of pennies on each side of the bowl. The process is not time-symmetrical because the reverse process, in which all the pennies end up on one side of the bowl, will never occur."

You take the bowl from Constantia and place it beside her borsht. "Actually, I shouldn't use the word 'never' to describe the chances that the pennies will at some point all again be on the same half of the bowl. There's a *small* probability that all the pennies will end up on either one of the two sides of the bowl. Let's imagine that you place the bowl on a shaking machine and take a photo of the bowl every minute to see where the pennies are located. With just one penny, the odds on a state of low entropy are one in one. The pennies must be on one side or the other, so every observation you make will surely indicate that 'all the pennies' are on one side. With three pennies, the odds that you will see all the pennies on one side drops to one in four. This means, on average, you will see this happen one time in every four times you examine the bowl. If you look every minute, you'll have to wait, on average, four minutes until you see this happen. With ten balls, you'll have to wait, on average, 512 minutes! You can calculate the odds of

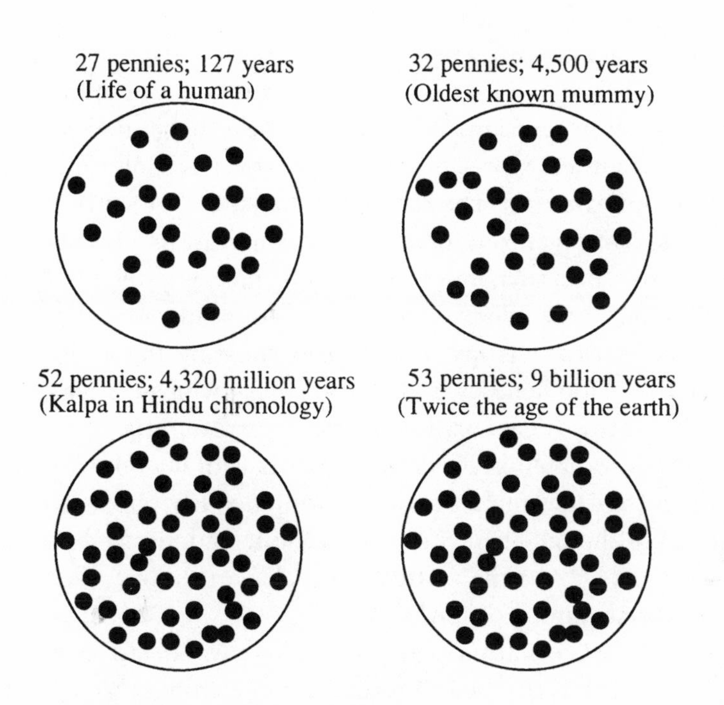

Figure 16.9 Time bowls for different spans of time.

finding all the pennies on one side of the bowl using $P = 2^{n-1}$ where n is the number of pennies."

You sketch on a napkin. "The following are some *time bowls* showing you how long you will have to wait to see all the pennies on one side of the bowl for different numbers of pennies. Twenty pennies in the bowl represent a year. That is, if you look at the bowl once a minute, you'll have to wait a year before seeing all the pennies in one half of the bowl."

You sketch a circle with 27 dots in it (Fig. 16.9). "Here is a bowl representing the life of a human (27 pennies corresponds to 127 years). The oldest known mummy is dated to about 4500 years ago. Here is a time bowl representing this period of time with 32 pennies." You pause and draw another bowl. "The longest measure of time is the *kalpa* in Hindu chronology. It is equivalent to 4,320 million years—roughly 52 pennies. Twice the age of the earth (4.5 billion years) corresponds to 53 pennies."

"All this helps to prove that, while it's possible that entropy (and therefore, in a certain sense, time) will reverse itself by chance, it is highly unlikely so long as the system involved contains more than a few elements. Let me explain. Most scientists believe that the second law of thermodynamics expresses the one-way nature of time for large macroscopic systems. The entropy of a system, which is related to a gain or loss of organization, must always increase as a result of the second law of thermodynamics. You are an *entropy-losing machine,* since you maintain a complex organization of atoms in your body. However, you achieve this entropy loss by disorganizing the atoms in food. (Much of the food energy is lost as body heat which dissipates into the air.) The entire *system* of you and your surroundings gains entropy. This one-way direction for entropy carries with it a one-way direction for time. Ice cubes melt, but don't spontaneously reform from liquid water. Even though elementary particle interactions can be symmetrical with respect to time, in our full-scale world, time, like an arrow, moves in only one direction."[3]

You follow Constantia's eyes, which have wandered to the window sill upon which is perched a bronze bust of astronaut Neil Armstrong. On the adjacent wall is a picture of John F. Kennedy and Nikita Khrushchev boarding a flying saucer.

Constantia's eyes are wide as she studies the interior of the restaurant. "Am I crazy, or does this restaurant decor seem a little out of place?"

You and Mr. Veil nod. Your eyes focus on the mounted head of a reindeer that is perched above the doors to the kitchen. The counter is made of oak, soaked through with lacquers and oil. There are various small wooden caribou statues on the shelves. Small Native American rugs lie on the wooden floor, before the fireplace and windows. Against one wall is a large aquarium filled with neon tetras, little red and blue fish about an inch in length. As they swim back and forth, they remind you of rubies, or perhaps the wisps and eddies of scarlet confetti. On top of the huge tank is a tiny plastic jar of Tetramin fish food. You recognize the brand, because it is the same one you use for your own aquarium.

You stand up and look around the restaurant. A woman with long blonde hair is eating a leg of lamb next to her young daughter by the counter. She holds the huge leg with both hands as it drips juices onto the table.

The mother sees your intent stare and looks back at you with an angry or nervous expression. It is hard to read her expression from faraway. "Hey, what are you staring at?" she says.

You look away as your eyes continue to dart around the room. Above the fireplace is a photograph of an astronaut. Under it, in big bold letters are the words, "Yuri Gagarin about to orbit the Earth (April 12, 1961)." Another photo shows a small gray alien with a bulbous head. Under the alien are the words, "Groom Lake, Nevada, 1999."

You see that Constantia and Mr. Veil are also staring at the photograph. "Feels like we're in a time warp," Constantia says. "I can't figure out if this is the craziest or most interesting restaurant I've ever been in."

You look at other people in the restaurant. Two men sit with a woman in a booth a few tables away. One of the men looks exactly like Mikhail Gorbachev, and he is talking to a young woman with shorts so short that most of her buttocks are on display. The other man appears to be Javier Perez de Cuellar, Secretary-General of the United Nations from 1982 to 1991. He is wearing a white shirt and tie. The woman with the shorts is smoking on a strange green weed.

Your mouth hangs open like a fish gasping for oxygen. "Will you look at that?" you whisper excitedly.

"Hey," Constantia says, "take your eyes off her." Constantia places her hand on your open lips and closes them.

You turn your face to Constantia. "No," you say, "it's Mikhail Gorbachev and—"

Constantia squints. "It does look like him." she starts.

The three of you stare at the bald man who is puffing on a smelly large brown cigar. His posture seems completely relaxed, but beneath this façade you think you see a trace of suspicion in the man's eyes.

"Sir," Mr. Veil says. "I think he's a Time Cop. Looks like Gorbachev but doesn't have the birthmark on his head."

Conversation at the other table stops. The Spaniard with the tie puts down his glass of rum. You recognize the label on the bottle. It is Ronrico, a fine premium spiced rum from Puerto Rico. The Spaniard's expression is fairly inscrutable.

You whisper back to Mr. Veil and Constantia. "Okay, stop looking at them. We're attracting too much attention by staring. Let's do one last experiment that helps demonstrate why time goes forward on a

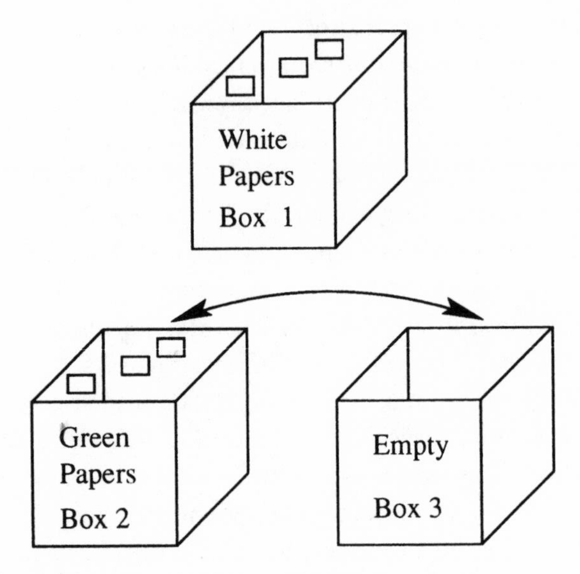

Figure 16.10 **Build your own time machine.**

macroscopic scale." You pull out three small boxes from your pocket and place them on the table (Fig. 16.10). "Box 1 contains 100 white slips of paper, numbered 1 through 100. Box 2 contains 100 slips of green paper, numbered 1 through 100. The third box, box 3, is empty." You look at Constantia. "Randomly choose white slips of paper from box 1. The number you pick will indicate to you which slip of green paper to move. For example, if you pick a white paper with the number 7 on it, then move the green slip of paper from box 2 to box 3. Replace the white paper in box 1, and repeat the process."

You are distracted by a fragile-looking waitress in an ermine loincloth, until Constantia coughs.

"Each time, the number on a white piece of paper tells you that a particular green piece of paper is moved either from box 2 to box 3 or vice versa. (If it is in box 2, move it to box 3. If it is in box 3, move it to box 2.) Can you guess what will happen after repeating this process 100 times?"

Constantia shakes her head as she begins shuffling papers.

"After some time has past, box 2 and box 3 will hold about equal quantities of green pieces of paper. You can think of the green pieces of paper as heat energy being homogenously distributed through space. On a macroscopic scale, energy flow and time flow proceed together—the direction of time. Entropy, which seems to give direction to time's arrow on a macroscopic scale, offers a *preferred* direction. While it's possible that the green slips of paper can all go back to box 2, it is very unlikely."

Mr. Veil waves his hand. "Is this all about the Second Law of Thermodynamics?"

You nod. "The Second Law of Thermodynamics suggests that as time runs forward, hot things get cooler and energy tends to become evenly distributed in the universe. However, the Second Law is not the same as most laws of nature. It's a statistical principle that works on the probabilities of what trillions of atomic and subatomic particles will do. In effect, it says that a hot iron will very probably get cooler, but not certainly. Individual particles are not subject to statistics. On an atomic level, it is quite possible that actual time reversals do take place, and that individual particles can move backward in time."[4]

In a few minutes, Constantia is sipping on a huge goblet of iced tea on which a sprig of crushed mint floats. You have an Andromedean beer made from the secretions of a two-headed crustacean that lived in an alien sea.

You turn to Mr. Veil. "What are you waiting for?"

"They are preparing my latest dish."

"Uh oh. I don't like the sound of it."

"Rafflesia, perhaps you've heard of it?"

"Rafflesia! That's the smelliest flower in the world." In fact, it is one of the strangest flowers on Earth. It mimics the texture and smell of decaying flesh to attract the flies it needs for pollination, Some call it the stinking corpse lily. You can smell it from sixty feet away.

Constantia gets up from the table and rolls her eyes.

Mr. Veil studies your expression for a few seconds, and says, "Sir, just kidding! I don't eat rafflesia."

"Kidding? I thought Zetamorphs never kid."

"I'm trying to learn." ✍

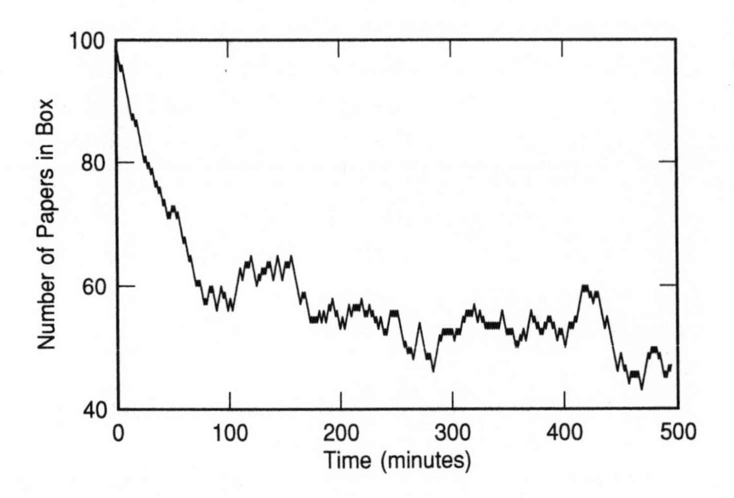

Figure 16.11 Results of the time machine experiment.

The Science Behind the Science Fiction

God seems to have left the receiver off the hook and time is running out.

—Arthur Koestler

You can build your own *entropy time machine* using your personal computer. The simulation just discussed involving white and green pieces of paper is outlined in Code 6 in Appendix 2. I like to think of the process as a cosmic hand (representing the laws of chance) moving balls of energy around the universe. Imagine that the cosmic hand moves the balls of energy once a minute. Figure 16.11 shows you the number of these balls of energy (or green slips of paper) that reside in box 2 as a function of time. We soon achieve a maximum state of disorder where about 50 percent of the slips of paper are in each box. How would the graph change if you started the simulation with ten times the number of paper slips or if you were to let the simulation continue for thousand times as long?

As we've also just discussed in this chapter, various chamber collections can be used to represent different spans of time. Why not try to draw time-in-a-chamber diagrams for the following ages? If you are a teacher, have the students draw each time span in three different ways by varying the number of backward connectors or the total number of chambers.[5] Which is larger, the number of possible chess games, which some have reported to be around $10^{10^{70.5}}$ or $C(15,20)$? Draw diagrams for the following (all times given in years from present): (1) Beginning of life on earth (3.25×10^9); (2) Age of fishes (3×10^8); (3) Age of mammals (7.5×10^7); (4) Age of mammoths (10^6); and (5) Isaac Newton (3×10^2).

Change an event in the past and you get a brand new future? Erase the conquest of Alexander by nudging a Neolithic pebble? Extirpate America by pulling up a shoot of Sumerian grain? Brother, that isn't the way it works at all! The space-time continuum's built of stubborn stuff, and change is anything but a chain reaction. Change the past and you start a wave of changes moving future-wards, but it damps out mighty fast. Haven't you ever heard of temporal reluctance, or of the Law of Conservation of Reality?

—Fritz Leiber, "Change War Series"

Backward time isn't such a new thing, backward time will start long ago.

—I. J. Good, attributed to Doog, J. I., 1965

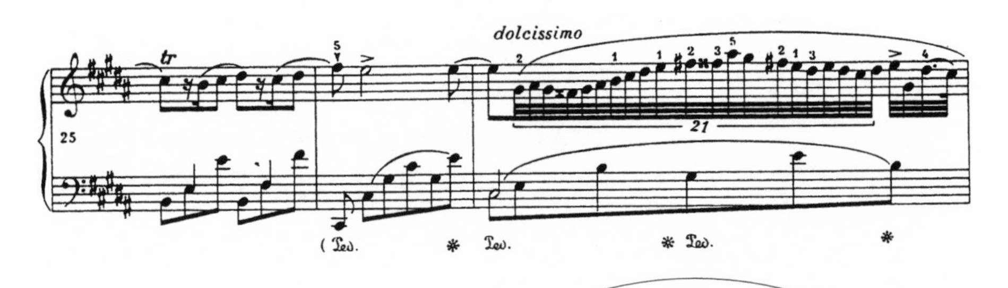

It's as if time were a rubber band, and you were on an end stretched out tight. If you pull the medallion off, we snap back to the future.

—Harlan Ellison, "Demon With a Glass Hand", *The Outer Limits*

Life survives in the chaos of the cosmos by picking order out of the winds. Death is certain, but life becomes possible by following patterns that lead like paths of firmer ground through the swamps of time. Cycles of light and dark, of heat and cold, of magnetism, radioactivity, and gravity all provide vital guides—and life learns to respond to even their most subtle signs. The emergence of a fruitfly is tuned by a spark lasting one thousandth of a second; the breeding of a bristle worm is coordinated on the ocean floor by a glimmer of light reflected from the moon. . . . Nothing happens in isolation. We breathe and bleed, we laugh and cry, we crash and die in time with cosmic cues.

—Lyall Watson, *Supernature*

Rotating Cylinders and the Possibility of Global Causality Violation

17

✍ "Today is the day. We're going back to meet Chopin!"

"Yes!" Constantia exclaims, giving you a hug.

"We'll meet him in 1829. Despite the lively musical life of Warsaw, Chopin urgently needed wider musical experience, and so his devoted parents found the money to send him off to Vienna. He made his performance debut there in 1829. A second concert confirmed his success—a success we'll soon witness!"

"Great," Constantia says. "I'm glad we put ourselves to bed last night with a hypnosleep course in Viennese culture and language."

You nod. Wienerisch, the Viennese speech and accent, reveals social levels and origins. The language evolved because the Viennese have been governed by Romans, Italians, Spanish, French, Magyars, and Slavs. The people in Vienna have absorbed Turkish and Yiddish words into their German tongue in a way that renders the original unrecognizable. The Viennese like their language because it proclaims their difference from their neighboring countries.

Vienna should be fun. If the people have any leaning toward pomposity, it is balanced by a habit of self-mockery, as expressed in their saying, "The situation is hopeless but not serious." The city's tourism and trade thrives on *Gemütlichkeit* (untranslatable but akin to "coziness"). The sentimental image of the Viennese—the nostalgia for wine, women, and song—is part of the city's allure.

You smile at Constantia. "Now don't be afraid if your stomach is a little upset for the first few days. I'd expect it to be a common accompaniment to time travel. It takes a little time for the good bacteria in your intestines to adapt to the germs of past centuries. In a few days, your intestinal flora will be back in working order helping with digestion and handling the new invading microbes."

"You don't make it sound very romantic."

"Nothing to fear. We'll bring the latest brand of Pepto Bismol—not the pink stuff but the rainbow-colored mix." You pause. "But first we have to finish our list of what we should take back." You bring out your list and read: "Video camera, film, can opener, underwear, gun, dynamite caps, and freeze-dried food sealed in nitrogen."

"Why the dynamite?"

"In case we get into some trouble with the locals. We'll try to blend in, but you can never be too careful. We'll bring some food in case the local stuff has horrendous bacteria."

Constantia hands you her list of items to take back.

"Wow," you say, "this is a long one. I'm impressed."

"Well, we don't know when or if we can come back to this time. For all we know, we'll crash-land in a nearby forest and have to hide from the locals who'll think we're invaders or witches. What if the car gets damaged? We might have to survive on our own until we can make repairs. Therefore, I want to be prepared."

You read her list out loud. "Tool kit, boots, scarves, gloves, handkerchiefs, lighters (or waterproof matches), notebooks, pencils, watches, compasses, spoons, penlights, field knives, space blankets, emergency rations, canteens, insect repellents, camouflage fatigues, medical kits, spare socks, toilet paper, individual camo netting."

"Good stuff," you say. "Do you have any hiking shoes?"

"No."

"Sneakers?"

"Yes."

"Good, you have ten minutes. Bring some heavy pants, some long-sleeved shirts, thick socks, and a jacket. None of your skintights. Try to avoid some of your more garish outfits. Past centuries may not be ready for them."

"Garish?" she says as she walks away to gather items.

You feel a tenseness and pop a diazepam cube into your mouth. You're uneasy because you aren't sure whether Constantia is in love with you. Perhaps you're worried about seeing Chopin play in the flesh in nineteenth-century Vienna. Would he live up to your hopes? Or will he be a disappointment? For the last decade you've cherished an image of ancient Vienna and of hearing Chopin in the flesh. Now on the verge of traveling there, you tremble.

You shake your head to relieve yourself of your anxiety. Vienna should be the ultimate adventure. Viennese *Lebenskunst* ("art of living") has survived changing rulers and times. Even in 2063, there are parts of Vienna where it's still possible to live at almost the same pace and in much the same style as it was a century ago. The same music is played in the same rebuilt concert halls, and a theatrical or operatic success still stimulates lively conversation. But today you can't drink the same sourish local wines in taverns on the outskirts of town, consume the same mountains of whipped cream at Sacher's and Demel's, nor sample the same infinite varieties of coffee in countless cafes.

You turn to one of your duffel bags and pack some thick woolen suits and overcoats in shades of green, gray, and brown loden cloth and colorful dirndl dresses—just in case you both need some more traditional garb.

In a few minutes, Mr. Veil and Constantia join you in the Museum's great hallway. "We'll use a Tipler cylinder to go back in time today. Do you recall our conversation about time travel in a Gödelian universe?"

"Yes," Constantia and Mr. Veil say in unison. Constantia continues. "A universe's rapid rotation produces a twist in spacetime and tips the light cones so that you can move in a closed timelike curve into your past."

"Right. But unfortunately, our universe doesn't seem to be Gödelian. However, we can produce the light cone tipping using a rotating, infinite cylinder. In this way we can create closed timelike curves. Let me show you." You sketch on a pad (Fig. 17.1). "The idea was proposed by physicist Frank Tipler in the 1970s." You pause. "A few

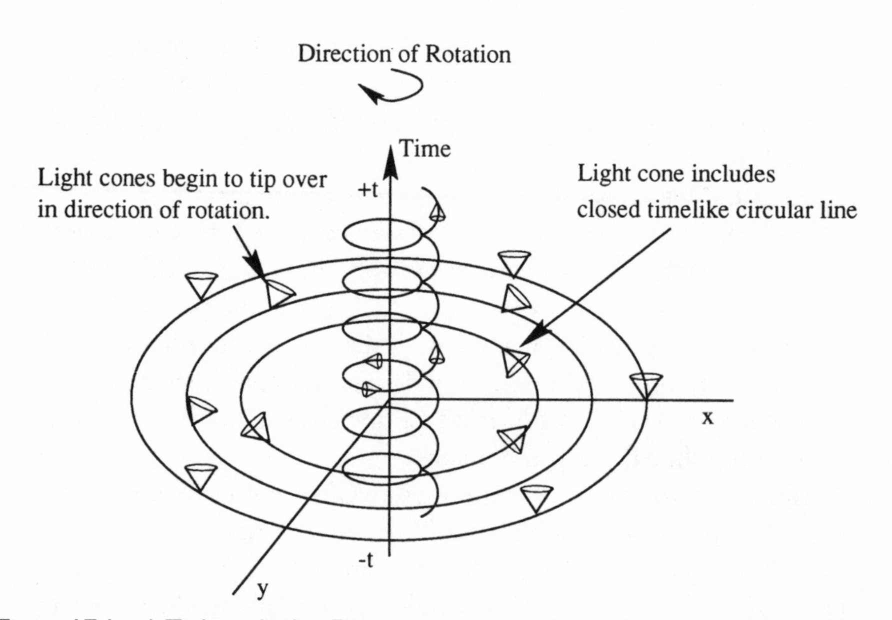

Figure 17.1 A Tipler cylinder. Future light cones point in the $+t$ direction far from the rotating cylindrical matter. However, as you get closer to the cylinder, spacetime becomes warped. The future light cones begin to tip in the direction of rotation. The helical timelike path allows one to move locally into the future in the $-t$ direction. This permits time travel to the past.

months ago I discovered such a cylinder spinning in outer space. I'll tell you about that later. Now, look at my diagram. The cylinder is aligned along the vertical axis. I've shown two space dimensions, *x* and *y*, with the flow of time, as usual, going up the page.

"The cylinder spins along the time axis, and its gravitational effects on light cones are possible to understand by looking at lines that trace circular paths around the cylinder at different distances. Far away from the spinning cylinder, the cylinder's warpage on spacetime is small, and the light cones are upright. This is the usual orientation for 'flat' spacetime. But as you get closer to the cylinder, the cones are tipped in the direction of rotation. I've illustrated this by drawing only the future light cones to keep the picture simple. Close to the cylinder, notice that the light cones are tipped in a direction that is spacelike far away from the cylinder. Near the cylinder the direction measures time, just as in the Gödelian universe. Therefore, to travel backward in time, all we have to do is approach the cylinder until we are near enough to be in the sufficiently tipped-over region of spacetime. Then Constantia and I will follow a helical path around the cylinder and spiral down in the negative time direction to the time of Chopin. We'll always be moving into our local future via the tipped light cones. Finally, we'll fly away from the cylinder, and then spacetime will become normal. Light cones will stand straight up. Then, we'll travel back to Earth."

Mr. Veil begins to pace. "Sounds dangerous, Sir. Are you going to travel at faster-than-light speeds around the cylinder?"

"No. The rules of special relativity restrict us to sublight speed. But to observers far away in flat spacetime, we'll be in a region where the roles of space and time interchange. Time itself begins to twist around the central cylinder. Time travel becomes possible when the light cones tip more than 45 degrees, at which point their edges lie below the x,y plane representing space in our diagram. Part of the future light cone in the region of strong gravitational fields will tip into the past as viewed from the region of a weak field. In fact, we could choose to move along a path that, to outside observers, consists solely of a circle in space, without any motion through time. We'd stay in the same space plane in the diagram, and we would appear to be everywhere around the orbit at the same time—omnipresent!"

Constantia claps her hands together. "Let's continue packing."

You nod. "We'll find plenty of food back in time, but let's bring something fresh with us in case we get hungry in the car."

Constantia opens the refrigerator in the museum's kitchen and starts piling whatever food she can find into a large bag. Polish sausage, mayonnaise, sweet pickles, olives, mustard, ketchup, a casserole of macaroni and cheese, sharp cheddar cheese, four tomatoes, and a loaf of rye bread. She grabs some forks and knives. "We're ready!" she says.

"Great. Go back and get the medical kit while I get the car."

You go through a back door in the museum to an underground garage. The hood of your sports car looks like a teardrop in the fluorescent light. The seats, within a transparent canopy, are made of leather.

You punch controls and lean back. The car whirls up a ramp and into the air.

It feels like autumn. Cold! The sports car is wet with dew. While waiting for Constantia, you check the engine, the rear thrusters, and the seals on the door. A little slack there, and you get out a tool kit and tighten it up.

In a few minutes Constantia is beside you. "Glad I dressed warmly."

"Let's go."

Mr. Veil smiles and bangs on the roof of the car. "Be careful now, okay?"

"We will," says Constantia. "Don't worry."

You turn to Mr. Veil. "Keep the museum running smoothly. We'll send you a postcard."

He smiles with multiple rows of teeth glistening in the receding sunlight.

You press the accelerator pedal. Constantia looks out the rear window until you are through the traffic lights at the far end of Fifth Avenue. In thirty seconds, Mr. Veil and the museum disappear from view behind rows of buildings.

By the time you are high into the air, the moon starts to shine above an ink-black Hudson River. You look down at the Woolworth Tower and a bright skeleton of the Brooklyn Bridge.

You turn to Constantia, who is comfortably seated beside you. "The cylinder is somewhere beyond planet Neptune. So sit back and relax. We're not sure how the cylinder got there, but it's made of an extremely dense matter to create the necessary amount of mass that sets space-time spinning. The material seems to have come from neutron stars—very dense objects formed when stars of a certain mass have undergone gravitational collapse. The stars are made of an extraordinary material called *neutronium.* Neutronium is so dense that a chunk the size of a thimble would weigh about 100 million tons. The sun, if squashed into a neutron star, would be only a few miles across, the earth just a few inches. The density of neutronium is 10^{14} times greater than ordinary solid matter."

Constantia runs her hands along the leather seats of the car. "Could an alien civilization have made the cylinder?"

"No one knows. Maybe they were able to herd neutron stars together using nuclear bombs. If they could find enough stars and fuse them together into a long cylinder, that would be half the battle won. This cylinder requires 100 neutron stars fused together, each about 20 kilometers (or around 12 1/2 miles) in radius, making a fused neutron star cylinder 40 kilometers across and 4000 kilometers long, about the width of the United States of America."

Constantia looks at the various gear in the back of the car. "What's all that extra stuff?"

"I have a small refrigerator and some hunting rifles in the back. I also got some detailed maps and satellite images of Vienna courtesy of National Geographic and the Smithsonian."

"Rifles?"

"As I said, just in case the natives aren't friendly."

Although you will do your best to make yourselves unobtrusive, you can never tell what someone might think of your odd ways. You wouldn't want some religious fanatic to burn you both alive for being a witch and warlock. Just in case, you have brought along .22 rimfires. Given Constantia's inexperience at shooting, the bolt-action .22s that have to be manually cocked after each round is chambered are the safest choice. No sense in killing people by accident. No sense in really upsetting the time stream.

"Ready to hibernate?" you say.

"All set."

You press a button on the car's front panel to let a gaseous mixture of melatonin and flunitrazepam fill the air. The last thing you remember is Constantia holding your hand as Chopin's *Fantaisie in F Minor* pours from the speaker.

When you awaken, planet Neptune is on your right. It has no apparent solid surface and its atmosphere contains methane gas. Absorption of red light by the atmospheric methane is responsible for Neptune's beautiful deep blue color.

Constantia stares at Neptune for a while. "Amazing," she says. She turns back to you. "Before we slept you said 'half the battle' is won by fusing stars into the cylinder. What did you mean?"

"Right, the second problem the aliens must have faced was how to align the positions of the stars so that the whole cylinder rotated as one unit. Typical neutron stars rotate at speeds between 1000 and 10,000 revolutions per second. They must have herded together the faster rotating stars, the ones that rotate at 10,000 rotations per second." You pause as Neptune recedes into the distance. "Constantia, can you believe it? With this rotational speed, the cylinder's surface probably reaches three-quarters the speed of light."

"You seem nervous."

"We've got to be careful. We don't want to get too near the cylinder where the gravitational forces are something awful. Instead we want to circle around it at a safe enough distance to avoid being torn apart by the gravitational tides generated by the cylinder, while at the same time being sufficiently close to use the spacetime warp surrounding the cylinder."

"Something else is bothering you."

You nod.

"Let's hope this cylinder has been around for many centuries. I'm not sure what would happen if we tried to go back in time prior to the creation of the cylinder."

Constantia points to some queer reflections of light. "What's over there?" she says.

"Let's take a look." You feel a bit nervous in the enclosed space haunted by the strange reflections of light. You look at Constantia, and she smiles at you to reassure you.

"There's the cylinder!" she yells.

The car headlights point toward it. The headlights flicker off for a second and then on again. You give an extra push to the button controlling them.

"Let's go closer, shall we?"

But Constantia does not answer. She looks up.

"What's wrong?"

"Oh my God," she says.

You try to turn your head, but she is blocking your view.

"What is it?"

She gasps.

You gently push her aside so that you can see. You bring your hand to your mouth, and for a moment you close your eyes.

It is an alien. Floating in space. Crushed. Its thin legs attached to its torso by mere threads. You guide the car away from it. "Poor fellow must have taken a wrong turn and gotten crushed somehow."

You turn the car toward the cylinder, when suddenly you see a red pulsing light to your right. It is long and slender with a glowing, red center. It moves through space and vanishes fifteen feet in front of you.

Your mouth hangs open. "What the hell?"

"Bozhe moi!" Constantia screams—a common Russian exclamation. Two glowing ellipsoids of spinning flame shoot by horizontally and vanish behind you.

You blink and try to steady the car. You're generally a pretty fearless person. Sure, when you were a child you had a few irrational fears—

fear of dark closets, fear of failing in school. But as an adult, you fear little. When you hear a strange, creaking noise in the museum, you are always the first to investigate. A rustling noise in the attic? No problem, you would be up there in a jiffy.

But now it is different. It is too real. A sudden lance of fear drives through you, an instinctive dread of the unknown. You quickly guide the car to the rotating cylinder in space.

"What the hell's going on?" Constantia says.

If you're dreaming, right now would be the time to wake up.

You look all around for a trace of the oblong lights. "I don't know. I think it's the Time Cops trying to scare us away from the cylinder." You softly curse when the car shudders.

Constantia is thrown forward. "Where's the closest safe place we can head to?"

"We have to head toward the cylinder and escape back in time."

The car starts shooting toward the cylinder at a strange angle.

"You're going the wrong way," Constantia shouts.

"Tell my car that," you yell back as you disable the autopilot and grab the manual steering wheel. The sky seems to get darker, and dust obscures your view. "I can't see a damned thing," you say.

The sky is suddenly filled with twirling flames, and then the burning ellipsoids spread out around your car.

You slam your foot on the accelerator, and the lights start to follow your car.

"To the cylinder," you scream, your eyes blinded by pulsing red lights. "What the hell's happening?" The car doors are shaking violently. Your stomach knots, and the air in the car suddenly feels very cold.

"We're getting close," Constantia yells. "Keep going." She looks back. "Those things are still following us," she screams as the colored lights roar above you. "What are they?" she wails.

You look for the ellipsoids of light, see nothing, rub your eyes. You can't believe this is happening.

You look up. "They're right next to me, damn it. I can almost hit them! Spinning cubes! Lights."

"Uzhas," Constantia screams in Russian, "more are coming!" A roaring sound starts. Subsonic booms seem to tear the very fabric of the sky from every angle, cracking chaotically through the heavens.

Suddenly a large object towers in the distance like a great cathedral made of glass and colored mirrors. Rows of yellow and blue lights pulse on its edges. You imagine its massive engines hummingly loudly, engines that sound strong enough to take it to the stars.

You press a red button on the car's dashboard, and start shooting projectiles at the lights. They disappear in an instant.

"Damn," you say shaking your head, flicking your eyes all around you. Your rational mind tells you that this is some kind of hallucination or trick. The whole gamut of possible explanations flash through your mind, but none seem to fit. You have stumbled onto something you know very little about. You can't explain what's happening, but you can't deny the lights that have been all around you. It must be either the Time Cops or some alien species that doesn't want you to use the cylinder for time travel.

"Don't shoot," Constantia yells. "It will make them madder."

"Them?" you scream. "It was just lights."

Constantia slumps backward as you shoot the car upward. She lies against the seat shaking a moment, and then she pulls herself up.

The lights appear again. "No!" you say, "now they're rising up." A bright white hole opens and closes on the lights' surface, and you hear an inhuman, whistling wail.

Constantia holds you, and you feel the rapid rise and fall of her chest, which is pressed against your side. She jerks her head around and starts shrieking, "What's that?" Her voice is nearly unrecognizable. You turn and see a pinwheel of violet lights heading toward you.

You are nearly crying. "This is too much," he said. "Are they trying to impress us with all these colors?"

Constantia looks back. "They're too fast for us to outrun."

You gaze all around, close your eyes, look again, see the violet pinwheel growing larger. The car starts making demented jerking motions. "Damn!" you yell.

"Almost there," Constantia screams, looking at the Tipler cylinder.

"Faster," you say to your car as you see a second pinwheel, then a third. The shock jolts your body. The adrenaline rushes like a roaring river.

The car starts jumping around like it's dancing the Watusi. "This car ain't going to make it!" you start screaming.

As you approach the creamy-colored cylinder, you see various debris and old tilting spaceships. The whole area is a cemetery of space wrecks. Closer to the cylinder are some of the oldest ships, weathered to such a degree that their names are all but completely eroded way. You see it all, but don't have time to say a word. Here and there are a few rotting space coffins, and—oh my God—a few bleached bones are sticking up through the fractured material of the coffins. Is that an alien creature you see picking on a few pieces of hair and splinters of bone?

You grab hold of the steering wheel with your strong hands, point it toward the cylinder, and close your eyes for a second. You crouch low, urging the car to go faster although it seems to need no encouragement.

The alien cemetery speeds by. "The car can't keep up this pace much longer."

The alien cemetery comes to an abrupt end. The cylinder you long for still looks far off. The cars engines are hot. Would the car run until its engine ruptures?

Constantia taps you on the back. "They're heading away," she whispers. "The lights. Whatever they are. They're—they're going." Her voice sounds like a whisper from the grave.

Your heart rate slowly returns to normal. Constantia takes a deep breath. You look at your car's speedometer. "I think that they don't want to get any closer to the cylinder. They tried to scare us off."

You look all around you. No sign of the strange lights. "Okay, let's do this carefully. There are various zones around the cylinder. In one zone, time simply ceases to exist. If we enter certain regions in the wrong direction, it can cause us great damage. By maintaining a high enough centrifugal speed around the cylinder, we'll be able to deal with the crushing gravitational tidal forces." You press a button. "Put your seat back. The sports car is flattening itself a bit to minimize the crush caused by the difference between gravitational forces at one side of the ship and the other."

You press a button, and Chopin's *Sonata in B Minor* pours from the car's speakers as you near the cylinder. Constantia clutches your arm.

"My God, it's beautiful," she says, staring at a whirling dervish of colors.

"You're seeing positron–electron pairs burst in tiny explosions. We should now be in the region where light cone tipping allows us to travel back in time."

You press several buttons on the car's console. "We begin our helical path around the cylinder." The car travels in a large spiral round and round the cylinder's time axis.

Constantia peers out her window. "We've been here before. Looks the same."

"We keep returning to the same spatial location, but at earlier and earlier times. We could always come back by following a similar helical path forward in time and back to the future."

You take a deep breath. "Okay, the year should be 1829. Let's pull away from the cylinder and return to Earth. As we move away from the cylinder we should begin to experience time synchronization as our flow of time starts coinciding with the flow in regions of space far from the cylinder."

There you are! Nineteenth-century Vienna! Your landing gear locks into place with a thud beneath you. You are lining up for a final approach.

You fly between the foothills of the Alps and the Carpathians, where the Danube river has cut its course through the mountains. A town is situated alongside the river, most of it on the right bank.

Constantia looks out her window. Down below lies a rolling expanse of tidy fields and patches of forest, some sunlit, some in shadow. Dirt roads stretch away into the distance.

You press a button on the car's dash to turn on three high-resolution television cameras on the car's roof. "I want to document some of this." On the ceiling, three small green lights indicate that the cameras are on and recording.

A map of eighteenth-century Vienna is stretched over Constantia's knees, your destination marked by a circle drawn in magic marker. As you get closer to the ground, you see a wooden water tower, groves of trees, a herd of oxen being driven down the road, and a man setting light to a bonfire. Low buildings sparkle in the morning sunlight.

Constantia watches it all roll past, her gaze shifting from detail to detail. It is a new world, a fertile, burgeoning one. You feel yourself relaxing. As you slowly glide over a clearing in the forest, both you and Constantia burst into applause.

You go to a slightly higher altitude, and out the window, far below, you see a large city. Judging from your flight path, you suspect it is Vienna. You are struck by the geometric outline of streets and avenues, the planned and ordered configuration of what humans had built, even in these primitive times.

You place the car on autopilot so that it maintains a circular path. "Let's have a quick snack before we land. Then we'll be all ready to explore on a full stomach."

You eat a lunch of sandwiches in the car as you gaze out at the wide valley ahead. "We'll listen to Chopin play at the Valldemosa Lodge on an estate that formerly belonged to the Austrian Catholic Church."

You take the car off autopilot. Within minutes a river of deep green slips silently by, its banks invisible beneath the overhang of thick vegetation. At first you don't see any people except for a little girl in a dirty white dress swinging on a wooden wheel suspended from trees. A couple of women are hanging out clothes on a line. All of the women, even the little girl, are wearing crucifixes.

The car's stealth technology serves you well. As a result, your sports car probably appears to most people as a quickly moving shadow.

Down the road you see a traditional fiacre, a two-horse carriage driven by a bowler-hatted coachman. You point it out to Constantia. "Ready to get out?"

"You bet."

You direct the car into a tiny clearing in the adjacent woods. Next, you reach overhead and flip a switch causing you and Constantia to be bathed in strong ultraviolet light. "It kills germs," you say. "We want to reduce the chances of our infecting them with modern microbes on our skin and clothes."

You flick another button, the car's locks retract, and the doors slide open with a hiss. As you get out of the car, Constantia lingers a moment and gazes out at the view. Beyond the road to the west, the land slopes downward and flattens out into meadows and forests.

You place a few tree branches on the car. "Let's camouflage the car."

You aren't too worried because even if the car is discovered, there's no way anyone could enter it. Its locks respond only to your and Constantia's voices.

You stretch your legs. "This is the way to the lodge." You point to a walkway that goes through a small iron gate and down several tiers of stone steps. The terrain grows increasingly hilly and rocky. Large granite outcrops overhang the path.

After walking for a few minutes, you reach the lodge's grounds. Immediately, the beauty of the lodge touches you. You are surrounded by colorful pastures and orchards, the trees of which seem unusually healthy. Giant elm trees rise up every hundred feet or so throughout the pastures.

After another few yards, the path bends west and up a rise. At the top of the knoll is the lodge, a large Viennese-style building constructed of hewn timbers and gray stone. The structure appears to contain at least thirty rooms, and a large screened porch coveres the entire east wall. The yard around the lodge is marked by additional gigantic elms and contains beds of flowers and ferns. Groups of people talk idly on the porch and among the elms.

The property is clean and fresh and unlike almost anything you've seen in New York in the year 2063. Some people on the grounds are opening up concessions stands, saying "*Guten Morgen*" and talking about how cool it is.

As you walk closer, you see stonework-framed arched windows. To the west are four castle-like corner towers and a three-story porch, gray columns, and black iron railings. Old copper gutters cling to the base of the roofs.

You enter the lodge and see seats but no one is sitting. Numerous food vendors line the back end of the room. "We must have arrived in the hall too early for the concert."

"Let's walk around," you say.

The naked beamed ceilings, though never built for exposure, are nevertheless beautiful without plaster, their black wood level because all the carpentry of these years is done with meticulous pride. You imagine yourself living in such a place, not as Constantia would have approved, but in splendor, with miles of polished floors before you as you make your way each night into this great concert hall to hear the local musicians. You like this building. It flames into your mind. If only you could transport such a building to the twenty-first century, make it your abode, live within its safety and grandeur in some forgotten spot of New York City.

Could Constantia imagine living in such a place? Whose wishes would be fulfilled if you somehow offered her this home? Who are you to think such things? Why, you could live with Constantia like Beauty and the Beast. You laugh out loud.

"My God," you say pointing to a beautiful piano in the corner. "It's a Cristofori piano."

In the Museum of Music you have the only three surviving examples of Cristofori's pianos that date from the 1720s. Your pianos are damaged and unplayable, but you could tell from studying them that they use intermediate levers to act on the hammers, providing an enormous velocity advantage. The hammers fly upward toward the strings much faster than the front end of the keys descend under the pianist's fingers. This adds to the crispness and sensitivity of the Cristofori's action. In addition to his innovative mechanism, Cristofori also introduced a unique double-wall case that isolates the soundboard from the pull of the strings. The sound of his instruments is said to be reminiscent of the harpsichord. The dynamic range should be surprisingly wide.

Constantia points to two busts on the piano and gasps. One is of Constantia. The other is of you.

You want to let out a scream, but instead carefully approach the piano. Your heart rate accelerates.

Constantia turns to you. "What—what could it mean?!" she whispers.

"Wait a minute," you say. "I think our imaginations are acting up. These don't look exactly like us. I'm not sure who they are, but they're not us. Probably some local pianists."

Constantia seems unsure but slowly backs away so as not to attract too much attention.

As you look around, it seems as if all the citizens of Vienna are entering the lodge and packing the courtyards in anticipation of the music. Food sellers are busy; you smell grilled ham, baked lamb, and stalls laden with goat cheese, nuts, and unfamiliar fruits.

"I'm starved," Constantia whispers.

You smile and buy her a spitted lamb, paying for it with bright silver coins worth a fortune to a coin-collector in the twenty-first century. Another vendor sells you wine out of a large barrel, letting you drink it right from the ladle. Other vendors notice your money and try to sell you an assortment of foods: blackened hard-boiled eggs and bags of animal organs including goatheads and testicles. This is the real thing! Nineteenth-century Vienna. The potpourri of odors, the reek of sweat and garlic, the colors, the noise, the laughter . . .

You look around rather than talk, catching fragments of conversation among people who seem to know each other and who glance at you and Constantia because you are new.

"Foreigners?" asks a young woman who is selling beads. "Where from? Switzerland? Italy?" She is a Mediterranean beauty; her eyes are dark with long lashes, her lips full, and her nose aquiline.

"England," you say.

The woman eyes you with awe, as though you have claimed to come from a distant star. "England! Fantastic! To think you came all this way to our city."

The woman is enthusiastic, strong, and fearless. She probably wouldn't be the least bit scared if she found your car in the woods. But the idea is ridiculous. The woman exists in an older world where comets are considered punishments from God, and most people lived and died without ever reading a book. If this woman came upon your car with its smooth metallic walls and cold flashing lights, she'd probably either run away screaming or faint like a dying flower.

The air is heavy and fragrant. Snatches of obscene song float up from the tavern district. But suddenly, the people grow silent.

"It's him," the young woman whispers.

A shiver runs down your back as you hear the approaching footsteps of Chopin on the polished teak floor.

You take your seats as handsome Chopin sits down on the piano bench. He smiles at the audience, and, for a moment, seems to look directly into Constantia's eyes. As if to heighten the incredible tension, he ever-so-slowly brings his hands to the keyboard where they finally remain motionless, poised above the keys like a hovering hawk before it strikes its prey. Suddenly his right index finger quivers, and then he begins with his *Piano Concerto No. 1 in E Minor.*

Constantia is deep in concentration, her chest rising and falling, her eyes never blinking. Chopin's music, no matter what the setting, is instantly recognizable!

My God; it's true. His unique sense of lyricism and unparalleled melodic genius astounds you. The notes he plays are slightly different from what you are used to. Perhaps he is playing an early version of the piece or he is improvising slightly.

He looks at the audience for a moment as his right hand, seemingly autonomous from his body, executes a thick cluster of thirty-second notes causing the audience to gasp and writhe in their seats. No wonder his music influenced so many composers who followed, from Brahms to Debussy.

Your breath comes in short spurts now. You are shaking. It's all worth it. The music. Vienna. Constantia. Chopin, the man before you, is truly among the immortals of music. He looks into the secret places of the heart. He draws all the magical sonorities from the piano as he becomes one with the rising and falling of the keys.

"His fingertips—they're—they're so smooth," Constantia whispers as his fingers dance at the speed of light across the keys. He starts a trill: it is too long; the seconds turn to minutes, a tantalizing invitation for more.

Constantia begins to rock back and forth, and you can imagine that a painful yearning is building within her. Chopin's tender hands roam over the keyboard as his feet pump the pedals. "Fantastic," is all that Constantia can manage as she licks her lips.

"Quiet," you say. "Everyone will hear us."

Chopin reaches the highest note on the piano, rarely ever played, and starts to gently tickle it with both his pinkie and the back of his thumb. Constantia's murmurs of appreciation become more gutteral, but you squeeze her hand until she is perfectly silent.

"Don't stop," she whispers.

Chopin continues his tickling. Constantia's eyes are wide, as she breathes in short shallow breaths, oblivious of who would hear, or who would care. Then she starts to rock back and forth. You think her pupils may have momentarily dilated.

You can't stand it any longer. The music is affecting you too, making you drunk with pleasure. You lean into Constantia. Her hair has the sweet but musty smell of a garden in early autumn, after a long, luscious rain.

It is time to leave Chopin. Life must go on. The lodge at Valldemosa is wonderful, but you have all of space and time to explore.

The next day you spend time with Constantia at the Cafe Demel, a local coffeehouse with evergreen branches hanging over the entrance. The wine is accompanied with music played by a trio of instruments: a fiddle, accordion, and guitar. This coffeehouse has been a Viennese institution for three centuries. It first opened with an inventory of Turkish coffee beans, part of the booty from the Siege of Vienna in 1683. On this day, there are a variety of coffees and an assortment of supplements, such as cream or brandy, to choose from. The Viennese have turned the coffeehouse into a sort of second living room, where they not only drink their beverages and consume pastries but also read periodicals, play cards, and chat with friends.

Waitresses with dark eyes and olive skin move serenely about you, refilling your cups, bringing fruit, cold legs of lamb, bowls of rice, and olives en brochette. Constantia and you gorge on soups, grilled duck, asparagus, and hard-boiled eggs served in turquoise enameled cups.

After you return to your car and see that everything is all right, you sleep with Constantia beneath the huge twisted limbs of an oak. In the morning, you eat from a loaf of bread and pieces of dried apricots. There is still chlorinated water in your canteens.

The third night you stay in an inn overlooking the Danube river. On this exact location, a Sheraton Hotel will stand 200 years later. The inn is a sturdy wooden building with a dining hall on the ground floor and large bedrooms above. The dining room has a round bronze-inlaid ivory table that could seat two dozen people. The library bulges with books and scrolls. Somehow you expected the rooms to be devoid of beds and to be forced to sleep on straw on the floor, but the beds are vaguely recognizable with mattresses stuffed with goose feathers.

That night you do little sleeping. There is too much noise and merriment outside. The air you breath is fresh and fragrant. You look out the window and see cobbled roads running between green meadows. A modest carriage drawn by two horses passes by. In the distance, cattle graze, and large oxen work a mill.

How can you sleep knowing that glorious nineteenth-century Vienna lies just outside the inn's door? You want to explore with Constantia.

You look around your cozy bedroom. The artwork on the wall resembles the engravings of Durer and Rembrandt. There is a closeness to everything—with a colorful Bruegel painting sitting at the head of the bed. "This is beautiful," you whisper.

Constantia smiles and nods. "Wonderful."

You feel goosebumps rise along your arms as you look at her. "I—I wish I had met you earlier in my life." You stare into her lovely eyes, which luminesce in the shadows of the Bruegel painting.

"We're together now," she says as she presses herself against you. She hugs you, runs her fingers up through your hair.

It is as if an electrical current is passing from her body to yours. You pull Constantia to you and kiss her on the lips.

*

When you hit the air again it is still chilly, but not like it was. To your right, prominently situated in the center of Vienna, is St. Stephen's Cathedral, one of the chief Gothic buildings of Europe. You circle it several times before shooting away towards a double row of cypress trees that guard a little-traveled side road. A splendid villa appears to your left. It has a large central courtyard, elegant colonnaded walkways, and walls with frescoes.

Your car passes through a number of towns and gradually your pulse begins to slow. The exhilaration begins to transform from jubilant exaltation to serenity. The sun feels good now. The road and green prairie farmland rush by you and sparkle in the sunshine. And soon there is nothing but the beautiful warmth and speed. There are yellow and gold daisies in the grass in front of an old stone fence, a meadow with some horses, and far in the distance, a low rising of the land with something scintillating on it like a tilted light cone. Probably a silo of a farm.

"Where to?" Constantia says.

"Let's head south."

You and Constantia shoot upward a little higher into the sky, and the drone of the engine becomes heavier. You top a rise, and see a new spread of land ahead. Fences are rarer, and the greenness deepens. You cut the engine and glide for a while, heading south.

Constantia presses a button on the car's dashboard, and Chopin's *Fantaisie in F Minor* pours into the car. Outside, there are no fences now, and the sweep of the hills is great. Below you is a lake. "Freedom," you whisper, and then you smile as orange light reflects off spirals on tessellated water, challenging the azure of an endless sky. You want to say something to Constantia, something profound, something that adequately expresses your emotions.

Constantia has tears in her eyes, and she stretches our her arms and says, "I know. We can go anywhere. We have shattered time." She gives your hand a squeeze. You think she looks like a goddess in the glorious pink rays of the sun. Her hand feels warm.

You stop, for saying the truth aloud is unendurable. You know now why this tranquil life seems like an after-life or a dream, unreal. It is because you know in your heart that it can be repeated, refined, and relived with slight alterations and sadness avoided. You are omniscient, omnipresent, omnipotent. You will never experience pain or loneliness or horror again. All this lovely play of form and light and color on the sea and in the eyes of men is no more than that: a playing of illusions in spacetime.

"Let's go," she says as you get closer to the Adriatic sea. Together you gaze into clear, limitless water. The car turns and glides above bright shining swirls of mist. You are moving toward open sea.

The Science Behind the Science Fiction

> Making a real alteration in the time flow is a difficult thing. You have to do something big, like killing a monarch. Simply being here, I introduce tiny changes, but they are damped out by ten centuries of time, and no real changes result down the line.
>
> —Robert Silverberg, *Up the Line*

> There was a young couple named Bright
> Who could make love much faster than light.
> They started one day
> In the relative way,
> And came on the previous night.
>
> —Anonymous MIT student

Our journey ends with a Tipler cylinder, first described in 1974 by Frank Tipler, a mathematical physicist at Tulane University in Louisiana. In one of the boldest statements to appear in a respected physics journal, Frank Tipler described specific construction details for a time machine. This time machine did not require the entire universe to rotate, as did Gödel's model. It did not require the enlargement of microscopic wormholes using hypothetical exotic matter. In particular, Tipler showed that a very dense, infinitely long cylinder allows a closed timelike loop to connect any two events in spacetime—if the surface of the cylinder rotates faster than half the speed of light (so that the rotation speed is such that the centrifugal forces are balanced by gravitational attraction). Such a configuration could be used to tip light cones outside the cylinder in a way similar to Gödel's light-cone tipping. Other researchers have suggested that an infinite cylinder may not be needed and that a 10-to-1 ratio of cylinder length to radius may be enough to create a time machine. Using such a time machine, one could travel into the past, although not to a time earlier than the cylinder's creation. Moreover, the cylinder would allow people to return to the future in order to come home again.

Of course, there are some practical problems with such a cylinder. For one thing, the cylinder might collapse under its own internal gravitation pressure. For another thing, the high-speed rotation of the cylinder could cause ordinary matter to explode. Perhaps the cylinder could be made of some superdense material, and Tipler himself has suggested the possibility of speeding up the rotation of an existing star as an alternative approach to building an actual cylinder.

In his book *Spacewarps,* John Gribbin calculates that a 100-kilometer-long, 10-kilometer-radius cylinder with a mass equal to that of the Sun and rotating twice each millisecond would meet Tipler's criteria for a time machine. This rotation rate is close to some of the fast pulsars, implying that certain astronomical objects may already be functioning as time machines.

Could we create a Tipler cylinder from the material of an asteroid or a linear string of neutron stars? As I just suggested, it would be difficult to prevent a finite rotating cylinder from collapsing along its axis like an accordion. Although centrifugal force prevents the mass from collapsing radially, the cylinder's own rigidity is all that resists the gravitational attraction along the rotation axis. For bodies sufficiently heavy to act as time machines, no known form of matter can hold out against this massive cylinder's strong axial self-attraction. Probably, such objects would collapse and form black holes, as discussed in my book *Black Holes: A Traveler's Guide.*

Frank Tipler published his mathematical paper describing a workable time machine in the journal *Physical Review D* under the title "Rotating Cylinders and the Possibility of Global Causality Violation." Science-fiction writer Larry Niven was sufficiently impressed with the concept to use not just the idea but the title from Tipler's article for a short story that can be found in the collection *Convergent Series.* In his story, Niven describes an interstellar war in which one side discovers a massive cylinder in space, an ancient artifact of an alien civilization. The generals try to place the cylinder in motion to gain certain military advantages.

Although humans or other sentient lifeforms may always have difficulty in building a Tipler cylinder, the universe may contain objects that are natural time machines; nature may already have done the job human engineers would find so difficult. The mathematics says it is possible.

For those of you who find it hard to believe the universe could possibly create cylinder-like time machines, recall how little we know of the universe.

As noted in a 1995 issue of *Time* magazine, "Cosmology has only lately crossed the dividing line from theology into true science." Due in part to images received from the Hubble Space Telescope, astronomers are renewing interest in the big questions. What is the universe made of? What is the fate of the universe? How is the cosmos structured?

With each new discovery of a new structure of the universe, we can only wonder whether there are various natural time machines in the fabric of the universe. In recent years, there have been many baffling theories and discoveries. Here are just a few:

- In our universe exists a *Great Wall* consisting of a huge concentration of galaxies stretching across 500 million light-years of space.
- In our universe exists a *Great Attractor,* a mysterious mass pulling much of the local universe toward the constellations Hydra and Centaurus.
- There are *Great Voids* in our universe. These are regions of space where few galaxies can be found. The universe also seems to have a fractal nature with galaxies hanging together in clusters. These clusters form larger clusters (clusters of clusters). "Superclusters" are clusters of these clusters-of-clusters.
- *Inflation theory* continues to be an important theory describing the evolution of our universe. Inflation theory suggests that the universe expanded like a drunken balloon-blower's balloon while the universe was in its first second of life.
- The existence of *dark matter* also continues to be hypothesized. Dark matter consists of subatomic particles that may account for most of the universe's mass. We don't know of what dark matter is composed, but theories include: neutrinos (subatomic particles), WIMPs (weakly interacting massive particles), MACHOs (massive compact halo objects), or black holes.
- *Cosmic strings* and *cosmic textures* are hypothetical entities that distort the spacetime fabric.

The NUT Universe

Gödeliean universes and Tipler cylinders are perhaps the most famous methods of time travel via warped spacetime. However, in 1963, Ezra Newman, Theodore Unti, and Louis Tamburino discovered a new solution to Einstein's equations even more provocative than Gödel's. Unlike Gödel's universe, the Newman–Unti–Tamburino universe did not have to rotate to permit time travel or closed timelike curves; rather, it required a spacetime warpage similar to one surrounding a black hole. Their universe appears so strange that it was quickly called the NUT universe after the inventors' initials. In a NUT universe, when your spaceship travels 360 degrees around a black hole, you do not end up where you started. Instead you enter another universal sheet. Think of this universe as a spiral staircase. With each turn of the staircase, a traveler enters a new universe. What would life be like in such a bizarre universe?

Initially, physicists rejected the NUT solution in the same way they had rejected Goedel's universe. Since our universe does not appear to behave in the same way as a NUT universe, physicists dismissed the theory. Experimental observations did not fit the theory. However, the NUT solution does not seem to violate any known laws of physics. Other physicists have created different kinds of solutions to Einstein's basic equations that seem to theoretically permit closed timelike curves and time travel. Tipler's cylinder in the 1970s was just one solution. Others include rotating black holes. Michio Kaku in his book *Hyperspace* eloquently summarizes the state of affairs:

> Einstein's equations, in some sense, were like a Trojan horse. On the surface, the horse looks like a perfectly acceptable gift, giving us the observed bending of starlight under gravity and a compelling explanation for the origin of the universe. However, inside lurk all sorts of strange demons and goblins, which allow for the possibility of interstellar travel through wormholes and time travel. The price we had to pay for peering into the darkest secrets of the universe was the potential downfall of some of our most commonly held beliefs about our world—that its space is simply connected and its history is unalterable.

Some Concluding M

> Gödel proved that the world of pure mathematics is inexhaustible; no finite set of axioms and rules of inference can ever encompass the whole of mathematics; given any finite set of axioms, we can find meaningful mathematical questions which the axioms leave unanswered. I hope that an analogous situation exists in the physical world. If my view of the future is correct, it means that the world of physics and astronomy is also inexhaustible; no matter how far we go into the future, there will always be new things happening, new information coming in, new worlds to explore, a constantly expanding domain of life, consciousness, and memory.
>
> —Freeman J. Dyson, theoretical physicist

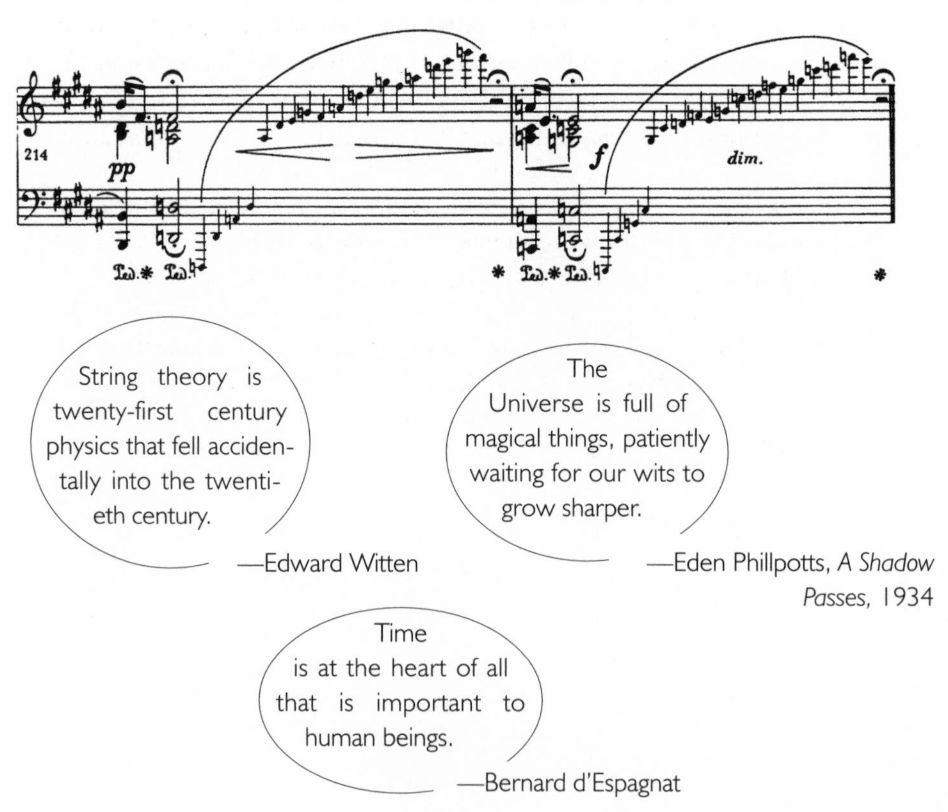

> String theory is twenty-first century physics that fell accidentally into the twentieth century.
>
> —Edward Witten

> The Universe is full of magical things, patiently waiting for our wits to grow sharper.
>
> —Eden Phillpotts, *A Shadow Passes,* 1934

> Time is at the heart of all that is important to human beings.
>
> —Bernard d'Espagnat

Some Concluding Musings and Thoughts

18

Spacetime

Ever since H. G. Wells's 1895 publication of *The Time Machine*—which described a four-dimensional spacetime with duration a dimension like height, width, and thickness—people have been wondering why we can not travel in time as we do in space. Einstein's special theory of relativity suggests that time is a dimension with similarities to the three spatial dimensions. In fact, the idea that time is a fourth dimension—on par with the familiar three dimensions of space—is one of the main foundations of modern physics (Fig. 18.1). If time is really just another kind of space, why can we not travel back and forth as easily as we move in space? If time is like space, then in some sense the past may literally still exist "back there" as surely as New York still exists even after I have left it. If we could travel in time as easily as we do in space, imagine how our lives would be transformed. We would no longer have regrets about past events. Nor would we wonder about "roads not taken." We would simply go back in time and make other choices and see what happens. If we were unhappy with the results, we would try again and again. Time travel would allow us to become omni-

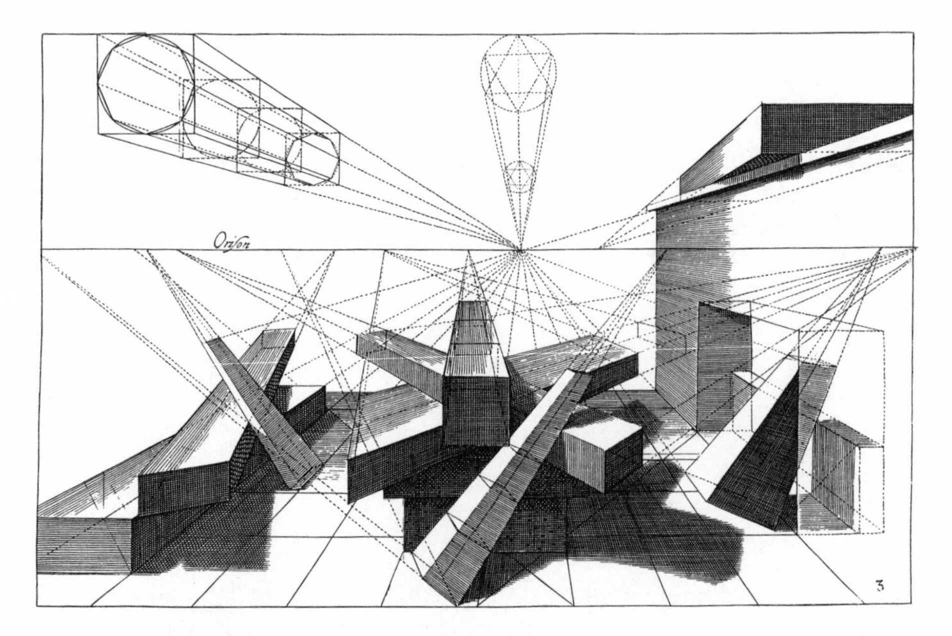

Figure 18.1 Modern physics suggests that time is a fourth dimension—on par with the familiar three dimensions of space. If time is really just another kind of space, why can we not travel back and forth as easily as we move in space? If we could travel in time as easily as we do in space, we would become different beings.

scient, omnipresent, and omnipotent—qualities humans have normally attributed to God. In a sense, time travel would allow us all to become God.

Some physicists believe that if we seriously consider time as a fourth dimension, then the past and future have always existed, and that human consciousness, for some unknown reason, perceives the universe one moment at a time, giving rise to the illusion of a continually changing present. As mathematical physicist Herman Weyl once noted, "The objective world simply is; it does not happen. Only to the gaze of my consciousness, crawling upward along the life line of my body, does a section of this world come to life as a fleeting image which continuously changes in time." Perhaps other beings in the universe do not have our perceptual constraints with regard to future and past, just like the creatures in Kurt Vonnegut's novel *Sirens of Titan.* In this book, the aliens see past and present at once and view human beings as "great millipedes with babies' legs at one end and old people's legs at the other." If time is truly a fourth dimension, then time travel always remains a possibility.

After having read this book, you should now realize that time travel is possible. For example, you have learned that high-speed travel warps time. Objects shrink in length, and past, present, and future become wildly mixed. Moving clocks do not remain synchronized with those standing still, and your moving body ages less rapidly when compared with your stationary twin. According to the theory of relativity, if you were to travel from our solar system to another solar system with a velocity close to that of light, events would proceed much slower for you then they would on Earth. (In *Star Trek,* such high speeds are common and allow the Enterprise to travel to the center of the galaxy in years, but sadly this should make it impossible for Captain Kirk to return home to see his relatives, because 90,000 years would have elapsed on Earth before Kirk came back.) It is theoretically possible to travel by machine as far as you wish into the future. In fact, if a spaceship could attain the speed of light (which does not seem to be possible because the theory of relativity tells us that the ship would have infinite mass), then time on the spaceship would stop. In Earth time, it might take 200 years for the spaceship to reach a destination, but to inhabitants of the spaceship, the destination would be reached virtually instantaneously. (This time stoppage reminds me of James Branch Cabell's *Juergen: A Comedy of Justice* where the main character stares into the eyes of God and is absolutely motionless for thirty-seven days!)

Many Worlds and Quantum Connections

In Chapter 5, we discussed Hugh Everett's "many worlds" interpretation of quantum theory, which suggests that a time traveler might travel to the past but in doing so he would be thrust into a parallel universe, thereby saving the original time line from paradox. This theory is quite congenial to time travel because it avoids all sorts of logical problems that might be created if you traveled back in time and killed your mother before you were born. (Sometimes I wonder whether there is a limitation or defect of our human consciousness that keeps us oblivious to different universes.)

We have also touched upon the concept of quantum connections—the notion that once two quantum systems have briefly interacted, they remain, in some sense, forever connected by an instantaneous link whose effects are undiminished through space no matter how far the particles travel apart from one another. Physicists' belief in the reality of quantum connection has been strengthened by Bell's theorem, which proves that superluminal links are required to explain certain simple quantum facts.[1]

Is quantum physics relevant to human experience of time? Many quantum physicists subscribe to the Copenhagen interpretation that suggests that despite the fact that the universe is quantum in nature, humans can never directly experience this quantum characteristic, and our experiences must always be "classical"—made up of discrete commonplace events. However, the quantum connection that instantly binds quantum particles might someday be used for FTL communication. The future will determine whether we can form a bridge between the quantum world and the world of ordinary human life. As Nick Herbert reminds us in *Faster Than Light,* "We may be sure that we have not discovered all the physics in the world; nature may contain more surprises in the area of time travel that conventional physicists have never dreamed of."

General Relativity

General relativity, Einstein's theory of gravity that involves curved space, enforces the speed-of-light limit locally but opens up the possibility of effective, long-distance FTL travel via twisted geometry (space warps). The Gödelian universe and Tipler cylinder arise from certain solutions to Einstein's gravity equations that describe space warps consisting of closed timelike loops. These loops permit time travel and communication through time. Because these solutions exist, we know that time travel is possible within the context of general relativity. Through history, physicists have found that if a phenomenon is not expressly forbidden, it usually occurs![2] Someday, computer simulations of such theoretical ideas may provide road maps for actual travel into the past.

Summing Up

Various researchers have proposed ways in which backward and forward time machines can be built that do not seem to violate any known laws of physics. Remember that the laws of physics tell us what is possible, not what is practical for humans at this point in time. The physics of time travel is still in its infancy. While all physicists today admit that time travel to the future is possible, many still believe time travel to the past will never easily be attainable. Don't believe anyone who tells you that humans will *never* have efficient technology for backward and forward time travel. Accurately predicting future technology is nearly impossible, and history is filled with underestimates of technology:

- "Heavier-than-air flying machines are impossible." (Lord Kelvin, president, Royal Society, 1895)
- "I think there is a world market for maybe five computers." (Thomas Watson, chairman of IBM, 1943)
- "There is no reason for any individual to have a computer in their home." (Ken Olsen, president, chairman and founder of Digital Equipment Corporation, 1977)
- "The telephone has too many shortcomings to be seriously considered as a means of communication. The device is inherently of no value to us." (Western Union internal memo, 1876)
- "Airplanes are interesting toys but of no military value." (Marshal Ferdinand Foch, French commander of Allied forces during the closing months of World War I, 1918)
- "The wireless music box has no imaginable commercial value. Who would pay for a message sent to nobody in particular?" (David Sarnoff's associates, in response to his urgings for investment in radio in the 1920s.)
- "Professor Goddard does not know the relation between action and reaction and the need to have something better than a vacuum against which to react. He seems to lack the basic knowledge ladled out daily in high schools." (*New York Times* editorial about Robert Goddard's revolutionary rocket work, 1921)
- "Who the hell wants to hear actors talk?" (Harry M. Warner, Warner Brothers, 1927)
- "Everything that can be invented has been invented." (Charles H. Duell, commissioner, U.S. Office of Patents, 1899)[3]

Wouldn't it be a wild world to live in if time travel devices played important roles in the development of humanity—like the computer and telephone? Mathematicians dating back to Georg Bernhard Riemann (1826–1866) have studied the properties of multiply connected spaces in which different regions of space and time are spliced together. Physicists, who once considered this an intellectual exercise for armchair speculation, are now seriously studying advanced branches of mathematics to create practical models of our universe.

Science-fiction stories about space travel have already inspired humans to travel to the moon. Similarly, will time-travel stories inspire us to create real time-travel mechanisms? Will we ever find a way to overcome the Einstein speed limit and make all of spacetime our home?

I wonder what humanity will discover about spacetime in the next century. Around four billion years ago, living creatures were nothing more than biochemical machines capable of self-reproduction. In a mere fraction of this time, humans evolved from creatures like *Australopithecus*. Today humans have wandered the Moon and have studied ideas ranging from general relativity to quantum cosmology. Who knows into what beings we will evolve? Who knows what intelligent machines we will create that will be our ultimate heirs? These creatures might survive virtually forever, and our ideas, hopes, and dreams carried with them. There is a strangeness to the cosmic symphony that may encompass time travel, higher dimensions, quantum superspace, and parallel universes—worlds that resemble our own and perhaps even occupy the same space as our own in some ghostly manner. Stephen Hawking has even proposed using wormholes to connect our universe with an *infinite* number of parallel universes. Edward Witten is working hard on superstring theory, which has already created a sensation in the world of physics because it can explain the nature of both matter and spacetime. By realizing that the fundamental laws of physics appear simpler in higher dimensions, string theory can unite Einstein's theory of gravity with quantum theory in ten dimensions. Our heirs, whatever or whoever they may be, will explore space and time to degrees we cannot currently fathom. They will create new melodies in the music of time. There are infinite harmonies to be explored.

Notes

Preface

1. At high speeds, objects also shrink in length. Moving clocks do not remain synchronized with those standing still, and your moving body ages less rapidly compared with your stationary twin. The notion of absolute cosmic time, with absolute simultaneity of events, was swept out of physics by Einstein's equations.

2. Of course, there is much debate about the specifics of time travel and what the universe will permit. For example, researchers such as Stephen Hawking have formulated the Chronology Projection Conjecture, which, if correct, would seem to rule out certain kinds of time travel altogether. However, as you read, continue to remind yourself that knowledge usually moves in an ever-expanding, upward-pointing funnel. From the rim, we look down and see previous knowledge from a new perspective as new theories are formed. Today's conjectures mutate, new theories evolve, and yesterday's impossibilities become part of everyday life.

3. Newton, whose father had died before he was born, was born on Christmas Day, 1642. In his early twenties, he invented calculus, proved that white light was a mixture of colors, explained the rainbow, built the first reflecting telescope, discovered the binomial theorem, introduced polar coordinates, and showed the force causing apples to fall is the same as the force that drives planetary motions and produces tides. Many of you probably don't realize that Newton was also a biblical fundamentalist, believing in the reality of angels, demons, and Satan. He believed in a literal interpretation of Genesis and believed the Earth to be only a few thousand years old. In fact, Newton spent much of his life trying to prove that the Old Testament is accurate history. One wonders how many more problems in physics Newton would have solved if he spent less time on his biblical studies.

Newton said much of his physics discoveries resulted from random playing, rather than directed and planned exploration. He once said he was like a little boy "playing on the seashore, and diverting myself now and then in finding a smoother pebble or a prettier shell than ordinary whilst the great ocean of truth lay all undiscovered before me." Newton, like other great scientific geniuses (Nikola Tesla, Oliver Heaviside, and many others), had a rather strange personality. For example, he had not the slightest interest in sex, never married, and almost never laughed (although he sometimes smiled). Newton suffered a massive mental breakdown, and some have conjectured that throughout his life he was a manic depressive with alternating moods of melancholy and happy activity. For more information on Newton, see M. Gardner, "Isaac Newton: Alchemist and Fundamentalist," *Skeptical Inquirer* 20, no. 5 (Sept./Oct. 1996): 13–16. For more information on the strange behaviors of great scientists, see C. Pickover, *Strange Brains and Genius* (New York: Plenum, 1998).

4. There are several excellent books on the nature of time, and these are listed in the "Reference" section. So, why another book on time? I have found that most books on time on the market today have a particular shortcoming. They are either totally descriptive, with no formulas with which readers can experiment —- not even simple formulas — or the books are so full of complicated looking equations that students, computer hobbyists, and educated laypeople are totally turned off. I don't shy away from formulas, but I give sufficient information so that readers can implement ideas with a hand calculator. My book, therefore, permits general readers to explore and readily understand time travel theories. The numerous figures are at the heart of much of the work described in this book. To understand what is around us, we need eyes to see it. Spacetime diagrams can be used to produce visual representations from myriad perspectives. In the same spirit as my previous books, *Time: A Traveler's Guide* combines old and new ideas — with emphasis on the fun that the creative person finds in doing, rather than in reading about the doing. Students may enjoy drawing spacetime diagrams to illustrate their own science-fiction stories or to gain a deeper understanding of time paradoxes. Seasoned researchers may enjoy the amalgamation of different topics all under one roof. Many of the chapters are brief and give you just a flavor of an application or method. Often, additional information can be found in the referenced publications.

Chapter 2

1. NIST-7 is one of the two most accurate timekeepers in the world and it resembles a ten-foot-long silver cannon. The clock sits in a National Institute of Standards and Technology (NIST) laboratory nestled at the foot of the Rocky Mountains, and is aligned due west. That way the sun moves across the sky parallel to the clock itself, and even the interaction of the solar radiation on Earth's magnetic field does not disturb the clock's accuracy. Since NIST scientists started building clocks in the 1950s (when the institute was known as the National Bureau of

Standards), the accuracy of their timepieces has improved steadily by a factor of ten every seven years or so. Among the most demanding users are astrophysicists, who are timing the phenomenally punctual pulsations of stars known as millisecond pulsars. NASA also needs the clocks to time the navigation commands it sends off to deep-space probes. People who work with telecommunications, the global positioning system, security, and defense also need to send or receive signals with billionth-of-a-second accuracy. Since 1967, the officially sanctioned length of a second has been defined by atomic standards: A second is equal to 9,192,631,770 oscillations of the radiation emitted or absorbed by atoms of cesium 133 when they undergo what is known as a hyperfine transition. For more information, see: G. Taubes, A Clock More Perfect Than Time, *Discover* 17, no. 12 (Dec. 1996): 69–71.

There are also attempts to build mercury clocks that will, in theory, be accurate to one second in 30 billion years! However, according to the general theory of relativity, the phenomenal accuracy of these clocks would be subject to small changes in the Earth's gravity resulting from elevation, tectonic activity, and changes in the Earth's mass resulting from particle radiation and meteor strikes. Therefore, two phenomenally accurate clocks will not indicate the same time if they are not exactly at the same location on Earth.

2. According to physicist Roger Penrose, the *visual appearance* of objects traveling at high speeds is not that of simple flattening suggested by the Lorentz transformation, which does not take into account the light travel time from different parts of the object to the observer. As a result, a sphere may look like a sphere at all speeds, while nonspherical objects can appear tilted.

Chapter 3

1. In the twenty-first century another genius will inevitably come along and present more comprehensive theories that will make Einstein take a back seat, just as Einstein superseded Newton.

2. To better understand a being who can exist outside of time, perhaps it is easier to first visualize a being who can move within a fourth spatial dimension. Theoni Pappas in *More Joy of Mathematics* discusses *hyperbeings* who can demonstrate the kinds of phenomena that occur in hyperspace. For example, a hyperbeing can effortlessly remove things before our very eyes, giving us the impression that the objects simply disappeared. This is like a three-dimensional creature's ability to remove a piece of dirt inside a circle drawn on a page without cutting the circle. The hyperbeing can also see inside any three-dimensional object or life form, and if necessary remove anything from inside. The being can look inside our intestines, or remove a tumor from our brain without ever cutting through the skin. A pair of gloves can be easily transformed into two left or two right gloves. And three-dimen-

sional knots fall apart in the hands of a hyperbeing, much as a two-dimensional knot (a loop of string lying on a plane) can easily be undone by a three-dimensional being simply by lifting the end of the loop up into the third dimension.

Chapter 4

1. Recent research on psychological repression gives further information on the brain's time machine. "Repression" is motivated forgetting of highly emotional or threatening memories. Attempts at suppressing thoughts about an event may slice a unified memory into static snapshots that often get recalled out of sequence, according to psychologist Daniel Wegner of the University of Virginia in Charlottesville. For more information, see "Out-of-Order Memories," *Science News,* 150, no. 19 (Nov. 1996): p. 301.

Chapter 5

1. Frank Tipler in *The Physics of Immortality* says, "If the universal resurrection is accomplished by reassembling the original atoms which made up the dead, would it not be logically impossible for God to resurrect cannibals? Every one of their atoms belongs to someone else."

Chapter 9

1. "Phase waves" can exceed the speed of light, but because these waves do not carry energy, this superluminal action corresponds to no real physical movement. Furthermore, phase waves are regular and predictable so cannot carry messages. Phase waves are similar to marquee lights in the sense that they can simulate an illusory sort of FTL motion.

Waves have various properties of possible interest to time travelers. For example, every kind of wave (e.g., water, sound, light, or quantum probability) obeys a certain wave equation. From a theoretical standpoint, Maxwell's wave equation for light has two solutions, the so-called "retarded solution" that describes a wave traveling forward in time and the "advanced solution" that describes a light wave traveling backward in time. Both waves travel at the speed of light in a vacuum. Ordinary light waves are called "retarded" because you always receive them after they are sent. Advanced waves, however, are received before they are sent. To date, no one has ever discovered advanced waves in any experiment. However, the appearance of advanced-wave solutions in ordinary wave equations demonstrates that the laws of physics theoretically permit backward-in-time wave motion.

Chapter 10

1. Some historical background: Black holes come in many shapes and sizes. Just a few weeks after Albert Einstein published his general relativity theory in 1915, German astronomer Karl Schwarzchild made exact calculations of what is now called the *Schwarzchild radius*. This radius defines a sphere surrounding a body of a particular mass. Within the sphere, gravity is so strong that light, matter, or any kind of signal cannot escape. In other words, anything that approaches closer than the Schwarzchild radius will become invisible and lost forever. For a mass equal to our sun's, the radius is a few kilometers. For a mass equal to the Earth's, the Schwarzchild radius defines a region of space the size of a walnut.

Chapter 11

1. To help you understand the incredible speeds of these accelerated electrons, I like to imagine a race between photons of light and accelerated electrons, such as those produced in Stanford's accelerator. In this race of particles, the electron can be made so speedy that it lags behind the photon by only 70 quadrillionths of a second.

2. In this chapter we have discussed antiparticles. You may be curious to find out that while creating hydrogen is as easy as mixing together electrons and protons so that electrons wind up orbiting protons, making antihydrogen is much more difficult. This is because the ingredients of antihydrogen—positrons (the positively charged antimatter counterparts of electrons) and antiprotons (the negatively charged antimatter counterparts of protons)—are more difficult to obtain, store, and control. Not until 1995 did physicists produce antihydrogen in the laboratory for the first time. Physicists would like to use antihydrogen to determine whether antimatter behaves in exactly the same way as ordinary matter. For more information, see I. Peterson, "Making Antihydrogen at Fermilab," *Science News,* 150, no. 22 (Nov. 1996): 340.

3. The ideas in this chapter come from various published scientific papers. The section describing negative energy of ultraluminal particles comes from Bilaniuk and Sudershan's articles and A. P. French's book *Special Relativity* (see References). The ideas on how the reinterpretation principle can lead to backward causation comes from Shoichi Yoshikawa of Princeton University (see References). There is a large literature on tachyon paradoxes (see, e.g., Savitt 1982 and Recami 1987). For a particle to go backward in time, its speed must not just be superluminal but the even faster ultraluminal. There has been some recent theoretical work that hints at the possibility of FTL photons! The effect is very small—well below any present experimental means of detection. See K. Scharnhorst, "On Propagation of Light in the Vacuum Between Plates," *Physics Letters B* 236 (Feb. 22, 1990): 354–359.

Chapter 12

1. Aharonov and colleagues derive a method to obtain a superposition of time evolutions of a quantum system that corresponds to different Hamiltonians as well as to different periods of time. They also consider its application to amplification of an effect due to the action of weak forces. A quantum time-translation machine based on the same principle, using gravitation fields, is also considered. In particular, they demonstrate that there exist superpositions of several time-evolution operators U_i that, for a large class of states $|\Psi>$, are effectively equal to a single but very different time-evolution operator $U' : \sum_i c_i U_i | \Psi> \sim U' |\Psi> t$. Their work provides an example of a new type of "time-translation" machine that is peculiar to quantum systems and has no classical analog.

2. In particular, Gott discovered exact solutions of Einstein's field equations for the general case of two moving straight cosmic strings that do not intersect. The solutions for parallel cosmic strings moving in opposite directions, each with $\gamma_s > (\sin 4\pi\mu)^{-1}$ in the laboratory frame, show closed timelike curves that circle the two strings as they pass, allowing observers to visit their own past. Similar results occur for nonparallel strings, and for masses in (2+1)-dimensional spacetime. For finite string loops, it is possible that black-hole formation prevents the formation of CTCs. For straight cosmic strings, the weak-field solution was followed by an exact solution whose exterior metric is given by $ds^2 = dr^2 + (1 - 4\mu)^2 r^2 d\phi^2 + dz^2 - dt^2$ where μ is the mass per unit length in geometrized units ($G = c = \bar{h} = 1$).

Chapter 14

1. The world line of the traveler is always timelike. This can be understood mathematically by taking the spacetime metric for Gödel's universe (with the standard convention $c = 1$): $(ds)^2 = (dt)^2 - (dr)^2 - (dy)^2 + \sinh^2 r(\sinh^2 r - 1)(d\phi)^2 + 2\sqrt{2}\sinh^2 r(d\phi)(dt)$ where t, r, y, and ϕ are cylindrical coordinates in four-dimensional spacetime, and by imagining the traveler's world line as the helical curve $r =$ constant, $y = 0$, and $t = -\alpha\phi$.

Chapter 15

1. Consider the true picture. Think of myriads of tiny bubbles, very sparsely scattered, rising through a vast black sea. We rule some of the bubbles. Of the waters we know nothing. . . .

—Larry Niven and Jerry Pournelle, *The Mote in God's Eye*

The quotation from Niven and Pournelle's futuristic science-fiction novel describes both the vast mystery of our universe and the strangely shaped objects we might one day encounter in outer space. In the 1990s, however, we do not need a space ship to explore strange new worlds consisting of bubble-like forms. Rather, all that is required is a small set of mathematical algorithms running on a good graphics computer.

Quantum foam schematic diagrams (known as "embedding diagrams" to physicists) will decorate living room walls of the future, and they will include computerized versions where the probabilistic froths are simply mathematical/computer graphical entities displayed on a computer screen. In order to create the undulating froth with wormholes, you may first wish to construct a two-dimensional model where the forms move, coalesce, and break up in the infinitely thin space between two glass plates. The simulation involves the use of "cellular automata" (CA). CA are a class of simple mathematical systems that are becoming important as models for a variety of physical processes. CA are mathematical idealizations of physical systems in which space and time are discrete. Usually CA consist of a grid of cells that can exist in two states, occupied or unoccupied. The occupancy of one cell is determined from a simple mathematical analysis of the occupancy of neighbor cells. One popular set or rules is set forth in what has become known as the game of "Life." Though the rules governing the creation of cellular automata are simple, the patterns they produce are very complicated and sometimes seem almost random, like a turbulent fluid flow or the output of a cryptographic system.

To create a CA, each cell of the array must be in one of the allowed states. The rules that determine how the states of its cells change with time are what determine the CA's behavior. There are an infinite number of possible cellular automata, each like a checkerboard world. Fig. 15.1 was produced by initially filling the CA array with random 1s and 0s. The rules of growth that determine the states of the cells in subsequent generations are discussed in the next paragraph. Graphically speaking, a "1" corresponds to a black dot, and a "0" corresponds to no dot.

The system in Fig. 15.1 evolves in discrete time according to a local law. As with most CA, the value taken by a cell at time $t + 1$ is determined by the values assumed at time t by the neighboring sites and by the considered site itself: $c_{i,j}^{t+1} = f(c^t_{(i,j)}, c^t_{(i+1,j+1)}, c^t_{(i-1,j-1)}, c^t_{(i-1,j+1)}, c^t_{(i+1,j-1)}, c^t_{(i+1,j)}, c^t_{(i-1,j)}, c^t_{(i,j+1)}, c^t_{(i,j-1)}$. In this equation, $c_{i,j}^t$ denotes the state occupied at time t by the site (i,j). The nine-cell template used in this chapter is referred to as the *Moore* neighborhood (as opposed to the *von Neumann* neighborhood consisting only of orthogonally adjacent neighbors). One interesting simulation simply examines the neighbor sites to determine whether the majority of neighbors are in state 1. If so, then the center site also becomes 1. We can represent those cells in the on (1) state as black dots on a graphics screen. In other words, this rule is a *voting rule* that assigns 0 or 1 according to the "popularity of these states in the neighborhood," and interestingly it generates behavior found in real physical systems. This simple *majority rule automata* produces hundreds of coalesced, convex-shaped black areas but does not lead to interesting graphical forms. A way to destabilize the interface between 1 and 0 areas is to modify the rules slightly so that a cell is on if the sum of the 1-sites in the Moore neighborhood is either 4, 6, 7, 8, or 9, otherwise the site is turned off. This rule has been studied previously (in lower resolution), and, since it uses a Moore neighborhood, it has been termed *M46789* by Gerard Vichniac in 1986. Such simulations have relevance to percolation and surface-tension studies of liquids.

The quantum foam in Fig. 15.1 shows an *M46789* two-dimensional form that has evolved after several hundred time steps from random initial conditions on a 2000 × 2000 square lattice. The tiny dust specks sparsely scattered between the coalescing blobs are not dirt left by the graphics printer, but rather they are stable structures such as

```
* *
* *
```

where each site has exactly 4 "on" members (out of 9) in its neighborhood, and thus stays on. The surrounding sites that are off have at most one or two neighbors that are on, and thus they stay off. There are probably quite a few other stable structures like this, though this rule does not seem to give rise to the zoo of stable objects allowed by, say, the game of Life.

Like a submarine pilot exploring coral formations in the Sargasso sea, computer graphics and powerful computers allow one to explore strange and colorful three-dimensional quantum froth tunnels and caverns using a mouse. As the simulation progresses, starting from randomly mixed 0s and 1s, those cells in the foam (that are in the 1 state) move around randomly until they meet and join by cohesion — forming visually interesting aggregates.

Chapter 16

1. This formula can be derived using concepts relating to random walk and diffusion problems. Depending upon whether or not you assume a continuous space or discrete space the answers are slightly different when the assemblage contains just a few chambers. An additional complication is that the first chamber is different from the rest. However, as stated, if a large number of chambers are considered, the time is proportional to n^2.

2. This equation, which may be used for $M \geq 1$, was derived in 1991 by Dr. Shriram Biyani using statistical arguments. A derivation of this formula is available from me upon request. It turns out that the *average* bottle number reached in a given time

approaches a constant. The formulas here give the average time until the *first* time the *n*th bottle is exited.

3. The neutral kaon is a special particle because it undergoes reactions that violate the law of time reversal invariance. For example, a time-reversed movie of $K_0 \rightarrow 2\pi$ would not look like a movie of the reaction $2\pi \rightarrow K_0$. Perhaps this reaction can be used by future time travelers as a kind of litmus test for assessing the direction of time's flow in unfamiliar universes.

4. A few scientists speculate that time will end. Austrian physicist Ludwig Boltzman long ago predicted the end of the universe as an attainment of maximum entropy. In this burned-out universe with no stars and no life, there will be no change by which time can be measured or observed. Some scientists have guessed that time will end on A.D. 10^{22} (10,000 billion years from now).

5. The oldest datable wine ever found was in two bottles in Xinyang, Hunan, China, from a tomb dated to 1300 B.C. Given only five time-chambers, how many backward connections would you have to use to represent the age of the wine? Here's another challenge. Although most astronomers believe the universe is around 15 billion years old, a minority have suggested much older ages such as 143 eons or gigayears (an eon or gigayear is one billion years). Represent this by time-in-a-chamber diagrams. The longest irrefutable age reported for any bird is 80 years, for a male cockatoo named Cocky who died in a London zoo. Represent this by time-in-a-chamber diagrams.

Time is frequently the subject of science-fiction movies, books, and television shows. In an episode of *Star Trek:The Next Generation* called "Time Squared" the starship *Enterprise* enters a time zone shaped like a Möbius strip. The characters were doomed to repeat the same sequence of events over and over again.

In 1990, the clock that maintained the primary time standard was called the "cesium NBS-6." The machine was six meters long. When operated as a clock the device keeps time with an error of about three millionths of a second per year. As we have discussed, in 1992, Hewlett-Packard unveiled an atomic clock that will remain reliable to the second for the next 1.6 million years. The $54,000 device is the size of a desktop computer.

In April 1990, *Playboy* magazine printed an interview with world-famous physicist Stephen Hawking. In the interview, Hawking speaks of "imaginary time." Here are some excerpts, to push your imagination beyond its breaking point:

> Imaginary time is another direction of time, one that is at right angles to ordinary, real time. We could get away from this one-dimensional, linelike behavior of time. . . . Ordinary time would be a derived concept we invent for psychological reasons. We invent ordinary time so that we can describe the universe as a succession of events in time, rather than as a static picture, like a surface map of the earth. . . . Time is just like another direction in space.

Chapter 18

1. However, some physicists suggest that an attempt to send FTL messages via Bell's theorem does not result in a *causal* connection (and is thus not in conflict with special relativity). Perhaps the harshest statements against superluminal links come from Victor Stenger, a professor of physics and astronomy at the University of Hawaii: "Nowhere does quantum mechanics imply that real matter or signals travel faster than light. In fact, superluminal signal propagation has been proven to be impossible in any theory consistent with conventional relativity and quantum mechanics." When commenting about Everett's many-worlds interpretation, Stenger adds, "Needless to say, the idea of parallel universes has attracted its own circle of enthusiastic proponents, in all universes presumably" (V. Stenger, "Quantum Quackery," *Skeptical Inquirer* 21, no. 1 (Jan./Feb. 1997): 37–40.)

2. Physics Professor Clint Sprott of the University of Wisconsin wrote me the following in response to this statement. His view provides another way of looking at this line of reasoning:

> The argument that everything happens that is not forbidden by the laws of physics is a powerful and useful one, but since the laws of physics are not yet all known, it is likely that we will someday have a law that asserts that such time travel is not possible. Meanwhile, I must concur that the possibility cannot be ruled out, and it is certainly fun to think about it and its implications.

Marilyn vos Savant is listed in the *Guinness Book of World Records* under "Highest IQ" and publishes an "Ask Marilyn" column in the Sunday Newspaper Magazine *Parade.* In the May 22, 1988, issue, Jennifer W. Webster of Slidell, La., asks, "What one discovery or event would prove all or most of modern scientific theory wrong?"

Marilyn replies, "Here's one of each. If the speed of light were discovered not to be a constant, modern scientific theory would be devastated. And if a divine creation could be proved to have occurred, modern scientists would be devastated."

I suspect that Marilyn is right about the speed of light. Einstein's special relativity theory, with its postulate that the speed of light in space is constant, is a linchpin that holds a wide range of modern physics theories together. Shatter this postulate, and modern physics becomes shaky in many areas.

3. Some say that Duell never uttered these exact words but that the general sentiment of the time was that most major technological advances had already occurred.

References

Aharonov, Y., J. Anandan, S. Popescu, and L. Vaidman. 1990. Superpositions of time evolutions of a quantum system and a quantum time-translation machine. *Physical Review Letters* 64 (June 18): 2965–68.

Allen, B. and J. Simson. 1992. Time travel on a string. *Nature* 357 (May 7): 19–21.

Bentov, I. 1977. *Stalking the wild pendulum.* New York: Dutton.

Bilaniuk, O. M P. and E. C. G. Sundarshan. 1969.Causality and space-like signals. *Nature* 223 (July 26): 386–87.

———. 1969. Particles beyond the light barrier. *Physics Today* 22 (May): 43–51.

Boulware, D. 1992. Quantum field theory in spaces with closed timelike curves. *Physical Review D* 46 (Nov. 15): 4421–41.

Cutler, C. 1992. Global structure in Gott's two-string spacetime. *Physical Review D* 45 (Jan. 15): 487–94.

Dyson, F. 1979. Time without end: Physics and biology in an open universe. *Reviews of Modern Physics* 51(3): 447–60.

Epstein, L. 1981. *Relativity visualized.* San Francisco: Insight Press.

Davies, P. 1995. *About time.* New York: Simon and Schuster.

Dennett, D., and M. Kinsbourne. 1992. Time and the observer: The where and when of consciousness in the brain. *Behavioral and Brain Sciences* 15: 183–247.

Dewdney, A. 1988. *The Planiverse.* New York: Poseidon.

Ehrlich, R. 1995. *What if you could unscramble an egg?* New Brunswick, N.J.: Rutgers University Press.

Everett, H. 1957. Relative state formulation ofquantum mechanics. *Reviews of Modern Physics* 29 (July): 454–62.

French, A. P. *Special relativity.* New York: W. W. Norton. Pp. 208–10.

Friedman, L., N. Papastamatiou, and J. Simon. 1992. Failure of unitarity for interacting fields on spacetimes with closed timelike curves. *Physical Review D* 46 (Nov. 15): 4456–69.

Gardner, M. 1970. Mathematical games: On altering the past, delaying the future and other ways of tampering with time. *Scientific American* 240(3) (March): 21-30.

———. 1992. *Fractal music, hypercards, and more.* New York: Freeman.

———. 1988. *Time travel and other mathematical bewilderments.* New York: Freeman.

Grabiner, J. 1974. Is mathematical truth time-dependent? *American Mathematics Monthly* 81: 354–65.

Gerrold, D. 1973. *The Man who folded himself.* New York: Random House.

Gödel, K. 1949. An example of a new type of cosmological solution of Einstein's field equations of gravitation. *Reviews of Modern Physics* 21 (July): 447–50.

Goudsmit, S., and R. Claiborne. 1966. *Time.* New York: Time-Life Books.

Gott, J. 1991. Closed timelike curves produced by pairs of moving cosmic strings: Exact solutions. *Physical Review Letters* 656 (March 4): 1126–29.

Herbert, N. 1988. *Faster than light.* New York: Plume.

Kaku, M. 1994. *Hyperspace.* New York: Anchor.

Krauss, L. 1995. *The Physics of Star Trek.* New York: Basic Books.

Mellor, H. 1981. *Real time.* New York: Cambridge University Press.

Morris, M., K. Thorne, and U. Yurtsever. 1988. Wormholes, time machines, and the weak energy condition. *Physical Review Letters* 61 (Sept. 26): 1446–49.

Nahin, P. 1993. *Time machines.* Woodbury, N. Y.: American Institute of Physics Press.

Nicholls, P. 1983. *The Science in science fiction.* New York: Knopf.

Niven, L. 1970. Rotating cylinders and the possibility of global causality violation. In *Convergent Series.* New York: Ballantine.

Nottale, L. 1991. The fractal structure of the quantum space-time. In A. Heck, and J. Perdang, *Applying fractals in astronomy.* New York: Springer.

Ori, A. 1991. Rapidly moving cosmic strings and chronology protection. *Physical Review D* 44 (Oct. 15): 2214–15.

Politzer, H. 1992. Simple quantum systems in spacetimes with closed timelike curves. *Physical Review D* 46 (Nov. 15): 4470–76.

Putnam, H. 1962. It ain't necessarily so. *Journal of Philosophy* 59 (Oct. 11): 658–71.

Recami, E. 1987. Tachyon kinematics and causality: A systematic thorough analysis of the tachyon causal paradoxes. *Foundations of Physics* 17 (March): 239–96.

Rucker, R. (1983) Jumping jack flash. In Rucker, ed., *The 57th Franz Kafka.* New York: Ace Books. (This story describes a shape-changing alien from a parallel universe. The alien uses a flatulence-based propulsion system to travel to black hole Gouda X-1. Once the alien dives through the black hole's ring singularity, he emerges in our universe where he masquerades as an English professor at Cornell University. The alien wants to repair a mishap suffered during a previous visit, and he plans his black hole transit so that he can travel backward in time, arriving moments before making his big mistake.)

Savitt, S. 1982. Tachyon signals, causal paradoxes, and the relativity of simultaneity. In P. Asquith and T. Nickles, eds., *Proceedings of the 1982 Biennial Meeting of the Philosophy of Science Associations.* East Lansing, Mich. Vol. 1, 277–92.

Silverberg, R. 1969. *Up the line.* New York: Ballantine.

Spear, G. 1992. Time travel redux. *Discover* April 13(4): 54–61.

Taylor, E., and J. Wheeler. 1992. *Spacetime physics.* New York: W. H. Freeman.

Tipler, F. 1974. Rotating cylinders and the possibility of global causality violation. *Physical Review D* 9 (April 15): 2203–6.

Visser, M. 1990. Wormholes, baby universes, and causality. *Physical Review D* 41 (Feb. 15): 1116–24.

———. 1993. From wormhole to time machine: Comments on Hawking's chronology protection conjecture. *Physical Review D* 47 (Jan. 15): 554–65.

Wolf, R. 1996. Believing what we see, hear, and touch: The delights and dangers of sensory illusions. *Skeptical Inquirer* 20(3) (May/June): 23–30.

Yoshikawa, S. 1969. Letter to editor. *Physics Today* 22 (Dec.): 47–52.

Appendix 1: The Grand Internet Time-Travel Survey

> Ask 100 theoretical physicists if faster-than-light speed is possible, and chances are, 99 of them will tell you no.
>
> —David Bauer, *Implosion*

> The mathematical spirit is a primordial human property that reveals itself whenever human beings live or material vestiges of former life exist.
>
> —Willi Hartner

The concept of time travel continues to fascinate both laypeople and seasoned physicists. Every one of us seems to harbor a secret wish to somehow relive the past and correct some small but vital mistake in our lives. Of course, if time travel were possible, then the laws of causality would be destroyed unless new, alternate universes are created when people travel back in time. There could be many problems. Just imagine a universe where, in 1963, scores of people suddenly arrived in Dallas to watch Kennedy's assassination. The streets would be crammed with people from the future bickering among themselves to see who would have the honor of preventing the President's murder. Consider another example. America's Desert Storm operation in the Persian Gulf would be botched as thousands of thrill-seeking Americans with cameras arrived taking pictures or as millions of Muslims arrived to jam military operations. In fact, history books could never be written! Some stubborn person would always be trying to assassinate Bill Clinton for his pro-choice support. Another crazed individual would kill President Lincoln earlier to delay freeing of the slaves. Yet another would kill Moses for freeing the Jews. If time travel were readily available, would our history be as ephemeral as the wisps and eddies of leaves in an autumnal wind?

Paul Nahin, author of *Time Machines*, aptly describes the wonder of time travel:

> A visit to the past is so mysterious and marvelously fascinating because it would let us watch ripples spread through time. Our one visit, in fact, might even be the pebble in the pond that starts an interesting ripple or two that will one day sweep over — us! Who would want to miss that? Indeed if modern philosophers are right, you can't (didn't/won't) miss it. I think time travel appeals, irresistibly, to the romantic in the soul of anyone who is human. For a time traveler passing back and forth through the ages, history would be the ultimate puzzle.

In March 1997, Roger Ebert of the *Chicago Sun Times* asked science-fiction visionary Arthur C. Clarke, "Do you think that there is any scientific invention or discovery so alarming that we would feel it necessary to suppress it?" Clarke's answer:

> There is one invention: a time probe, revealing the past and everything that has ever happened. When you think of the implications of that, it's pretty appalling. Could we as a human civilization survive such a thing? I just don't know. I don't think it's possible, but I can't rule out that a time probe could be created. All the mysteries and all the secrets that had ever happened would be revealed. Total transparency. That is the most terrifying invention I can think of.

In this appendix, I quote from the responses of physicists and laypeople around the world who have answered my questions regarding time travel, and I thank them for permission to reproduce excerpts from their comments.

Questions

Among the questions I asked were: Will time travel ever serve a useful purpose for humans or some other advanced civilization? What aspects of time travel do you find most fascinating? and What impact would a "time-viewer" have on society?

Answers

Craig Becker, computer programmer and inventor from Austin, Texas: . . . I believe that time travel to the past is impossible. On the other hand, I won't rule out the possibility of some kind of "time-viewer" that allows one to see into the past, because this avoids the possibility of paradox. I would refer the reader to the Damon Knight short story "I See You" for a nice, stark, treatment of the social consequences of something like this becoming popularly available.

Jim Glass, aerospace propulsion engineer and avid science-fiction reader from Canoga Park, California: [I]t's obvious that a "real" time-viewer would allow one to see anyone anywhere at anytime. You could use it to look inside your neighbor's house one millisecond in the past. It would probably destroy society or would be outlawed with draconian penalties for ownership/use.

Dan Winarski, computer programmer and inventor from Tucson, Arizona: The public is fascinated with time travel because this is viewed as an escape, but I believe that the most common misconception the public has about time travel is that one can go back in time. Theoretically one can travel forward in time by traveling at high speeds (which retards aging relative to stationary observers), and returning to an Earth that has aged for centuries. I believe, for whatever reason, that time is like a diode. You can go forward but you cannot go backwards. On a personal note, I find the time–velocity relationship most fascinating. Note that this relationship is positive definite, meaning that if you move at the speed of light in one direction then turn 180 degrees and move at the speed of light in the reverse direction, you merely reduce the progress of time to zero. Time does not flow backwards.

Brad Pokorny, former science-writer for the Boston Globe *and currently a journalist writing about international human rights and sustainable development as editor of* One Country, *a Baha'i newsletter. (The following views are his own and do not necessarily reflect the beliefs of any organization):* Time travel is not possible because it creates too many paradoxes such as being able to go back and kill your own grandfather, thus causing you to cease to exist, thus causing you not to be able to go back and kill your grandfather, thus causing you to re-exist, etc. I don't believe that God will allow this, ever. (For those of you who wish to avoid the concept of God in scientific discussions, you can substitute the word "Universe.") The nature of the universe is orderly and it will not allow time travel paradoxes. People of the Baha'i faith believe that in the next life, when we live a spiritual existence, we will be freed from the barriers of time. I do not know how this works, but it is my understanding that this would not create paradoxes, inasmuch as the spiritual world does not generally impinge on the physical world.

The [time] viewer would have little effect because events are subject to interpretation. People will believe what they want. Even if people, for example, viewed Christ rising from the grave, it would not change much. Those who don't want to believe would think it a trick, would doubt the viewer. Those who already believed would have their beliefs validated, but it would not really change people's behavior. This holds for other historical events. What does it matter if we get to see how the Battle of Gettysburg really progressed, or if we can see Cleopatra let an asp bite her, or find out whether Lee Harvey Oswald really had an accomplice? It is all academic. Some interesting papers might be published. But it would not have much of an effect on society.

Don Webb, author of the highly acclaimed science-fiction book Spell for the Fulfillment of Desire, *a world-renowned expert in the history of science fiction:* Time travel will not occur since memory of the future (on a non-quantum level) would violate laws of conservation of energy. However, people may develop ways to resonate with optimal future existences. If we can learn to "echo-locate" our optimal future, mankind will take a quantum leap forward. I think most people would be surprised that there actually are non-nutty people seriously looking at the physics of time travel. . . . Time travel would let us out of the box of three dimensions and five senses. It would let us out of the box called "Death." We would know most of the things we ache to know given mankind's longing spirit and limited existence.

I think a time-viewer would be more devastating that we can imagine. We would lose our myths and heroes, and we would turn those moments made great by our hearts and minds into entertainment.

Mike Hocker, computer programmer and inventor from Poughkeepsie, New York: I believe that time travel into the past is possible, but that the net result of intervention in the past is a future where the creator of the time travel machine does not exist; ergo, the feedback result is that no one discovers time travel.

Personally, I would like to have a viewer into the past. This would be far better than the Internet. . . . The time-viewer would blow away the revisionist historical accounts of past events. A lot of "heroes" would be found to have clay feet.

Dan Evens, nuclear design engineer and physicist who specializes in quantum field theory and gravity from Toronto, Canada: Time travel already serves a useful purpose. It provides a useful basis for endless late-night TV shows.

To understand a time-viewer's impact, see the movie *Millennium.*

Leonard Erickson, science-fiction aficionado and member of the Society for Creative Anachronism: As Isaac Asimov noted in a short story years ago, a time-viewer is the ultimate spying device. Since to see anything, it'd have to be "steerable" (i.e., you could use it to travel in space as well as time), and since the past starts a microsecond ago, I can watch anyplace in the world. If the time-viewer is somehow limited to travel in the same location (unreasonable, given that the earth, sun, and galaxy are all moving in different directions at high speeds), then it's still dangerous. Who needs detective work? Just set up a viewer at the scene of the crime and tune back.

Jennifer Kramer, freelance journalist: If a time-viewer were available, we'd spend even more of our days in front of one glowing screen or another.

Clark Alford, physicist: Yes, time travel will someday serve a useful purpose, despite the fact we cannot currently determine how this can be. The universe is infinitely richer than our small biologically based brains can imagine. We are just beginning to scientifically speculate about black holes, parallel universes, wormholes, string theory. . . . As we step into the future, with computers aiding our senses, we will find new worlds to explore, a constantly expanding domain of life and consciousness. Time travel of sorts is already known to be possible. Time travel will serve a useful purpose in the next millennium.

If a time-viewer were readily available, it would ruin the notion of privacy—which would not be good. I would use a viewer for all the reasons just stated, and I believe that most people would use it for sexual voyeurism, even though they would not admit their interest. If you gave the viewer to your friend, could he resist watching your wife and you together at night? On the good side, all sorts of paranormal, unskeptical, and religious beliefs would vanish in an instant, because humans would now be able to verify that there are no miracles, no UFOs, no phenomena unexplained by science. Creationism would die. Religious-based fanaticism and its evils would die. How could Ayatollah Khomeini send off children to die in war by offering them rewards in heaven if people can verify there are no prophets who performed miracles? Born-again Christians would have to look for solace elsewhere. The number of abortions would decline, because a woman could avoid those sex acts that lead to "unexpected" children. Let's change the subject for a moment. Seeing some of the horrors in the past may not be good for society. How much could we take of seeing the cruel rape of blacks by their white masters in the South, the gassing of Jews by Hitler, the horrors of the Spanish Inquisition, the routine torture and human sacrifices of the Aztecs and other pre-Hispanic Central-American cultures.... Hopefully, seeing these things would help us avoid similar incidents in the future. Interestingly, if it were possible, society would create an entire industry of time-traveler "shields" that one could drape over the bedroom so that time-travelers could not watch private acts. Perhaps certain kinds of carnal acts would be avoided by people, because they would fear being spied upon.

Gabriel Landini, research fellow in Oral Pathology at the University of Birmingham, U.K.: I have experienced what I consider a "virtual time travel." Let's suppose that you are playing a computer game in which you can "save" the status of the match for replay at a later time. When you load your game status back, you go to the past or reassume the present. If things go wrong in the game, you can also travel back to the point where you saved it and try to solve the past in a more convenient way. The more instances you saved the game, the more places in time you have to travel back. If things go wrong again, just travel back again until you succeed. If you go to the past and keep saving locations as you replay then it may be that you can have several possible "parallel pasts" to go back to, one (or more) for each saved game during your travel to the past or the future. Travel to the future is then possible only if that future has been lived before. While this works in software, in real life it would require freezing the status of the universe to re-create it later. Have you got a floppy disk with you?

A time machine would also function as a "transmatter" machine. This version is the one that governments allow the public to use. You just change the spatial coordinates and keep the time coordinate unchanged: travel in no time. There would be no more waiting in the traffic jams, but there would be large queues at the transmatter station. The transportation industry collapses; roads, airports, ports are empty; no more traffic accidents. Transmatter Inc. shares go up. Rapid exploration of the galaxy is possible. Extraterrestrial life is found.

In a sense, we have time-viewers. Old movies do just that. A time-viewer would be, of course, more convenient and precise. It would allow people to see exactly how events have occurred, such as the extinction of the dinosaurs. It would be used in courts to prove criminals guilty or confirm alibis of innocents. Changing subjects, what effect would a time-viewer have on quantum theory and the "Heisenberg uncertainty principle"? For example, assume we want to measure the spin of an electron, but normally we cannot also measure its position. No problem, we go back to the moment of measuring the spin and now measure the position. At the end of the measurements you know both. Therefore the only impediment to reconstructing all the past and predicting all the future arises from chaotic behavior, but not Heisenberg's principle. . . .

Morgan L. Owens, mathematics graduate from the University of Auckland: Oddly enough, the theoretical underpinnings for a time-viewer that focuses only on the past are more firm than viewers that focus on the future. Need a way to patch the ozone layer? Scope ahead to see how it will be done, and implement that scheme! What if you discover that it won't be patched and will collapse completely, and that our descendants will all be mutated out of recognition? What's the point of even trying, then, if such a bleak future has already been written . . . and we soon find ourselves on the oxymoronically horizontal slopes of Lake Paradox.

Peter Andrews, web page editor/producer and one-time chemist: Backward time travel is very difficult to implement, even if it were possible for particles. However, humanity has an eternity to solve this problem, and we will make it happen eventually. . . .

C. Delisle, university student: Einstein said it best: "It would seem appallingly clear that our technology has outgrown our humanity." As long as this is true, time travel is a danger to human society. Since the temporal events sequence can be upset by even a change in thought patterns, any interloper in the past has a chance of unintentionally interfering with time. This fits with chaos theory. For a more advanced civilization, time travel could be an incredible benefit if controlled properly. The understanding of past mistakes and successes would allow for faster development and a greater sense of the struggles that created the present. In addition, the ability to move anywhere in spacetime would give a race virtually unlimited exploration. (I use the term "spacetime" because one cannot move only in time or only in space since they are simply different aspects of the same thing.) An understanding of time itself is the first step to be taken. I would like to see this explored using current subatomic, quantum level techniques in the next fifty years. This might lead to the discovery of temporal subdimensions similar to the spatial dimensions of length, width, and depth.

John de Rivaz, editor of the popular newsletter Fractal Report *(U.K.):* If our universe is a many-worlds quantum universe, then I would use the [time] viewer to find a probability that suited me. If there is only a single time track, I'd be very careful and use it to make money in this time by finding lottery numbers and stock market trends. Let's assume that the time-viewer is a chronoscope, a device where the user inputs a time and space coordinate, and it then starts showing sound and visuals of events at that space and starting at that time. Perhaps the device could be used to speed time, like the fast play on a VCR. Note that the legal profession would hate this. Courts would be virtually redundant, as any crime can be viewed in its gory detail. If I discovered how to build a chronoscope, I would start by viewing events in my own life (even if only just to see it was working properly). As far as history is concerned, the choice is more difficult. Maybe the technology could include an automatic language translator. A translator is possible because if we discover the human brain's equivalent of machine code which should enable languages to be translated easily, otherwise the device would have limited use. Even 14th-century English would be incomprehensible to most.

Brian Sams, meteorologist and electrical engineer: I've always wondered why time travel devices are portrayed as big machines that transport one to their destination, but only a little box to get back. I think that time travel, at first, would be a one-way ticket unless the entire machine was sent.

Daniel Platt, computer programmer and theoretical physicist: Time travel is problematical. Let me give readers a detailed explanation which is probably of interest to physicists reading this book. Others should skip this paragraph. First consider that physics is permeated with the notion of conserved quantities. To a physicist, this means"you have the same stuff now as you did then." In terms of a four-space (relativistic) formulation, it just looks like you have a divergence-less flux vector. Given this, time appears similar to any other dimension, except that the relativistic metric is hyperbolic, with time being the odd-dimension out (sign different than all the others). As far as physics even giving you a direction for time, it only does so weakly. Entropy increase as a statistical phenomenon means that you tend to run the clock one way or the other, but you cannot reverse the clock in mid-stream. If you did reverse the clock, a movie would show occurrences of very improbable things. For example, if you start with all the molecules piled up in a corner of the room, then almost all possible trajectories will make the particles fill the room. If you reverse the velocities of all the particles with all the atoms in the corner of the room, for almost all combinations of initial velocities, the atoms will still fill the room. Even in the special case where the atoms want to pile up into an even smaller volume, it will take a tiny amount of time for the particles to reach a minimum compression, and then fill the whole room. Statistics doesn't say much about mechanical reversibility. There has been no measured discovery of any particle whose symmetry might violate this (aside from a few reactions that seem to violate CP symmetry). This implies either the particles prefer a time direction or they prefer left-handed coordinate systems over right-handed coordinate systems. There are no particles that seem to violate CPT symmetry, which would yield a direction to time. Our experience of time suggests a past (comprised of historically determined facts), the present that we are experiencing, and an as-of-yet undetermined future. Time is even built into our languages in the form of tense. Yet, capturing this very basic aspect of our experience has proven very elusive from the standpoint of physics. Instead, physics takes it as a given. Physics can be deceptive in that it tends to find all sorts of subtle connections between things. But when you face a problem like this, you realize physics is just description with lots of rigorous mathematical machinery forcing consistency between various parts of that description. There's only one place where time reversal seems to be a vague possibility. If you look at black holes, the metric inside the event horizon reverses spacelike and timelike coordinates. The radius starts to act timelike, and time starts to act space-like. Transformations exist that move across the event horizon smoothly (the event horizon is not an essential singularity in the same way that the center of a black hole would be). Those transformations, called Kruskel transformations, also point out a structure that admits completion in something called a "white hole." However, with a simple Schwarzschild singularity, you cannot get there from a normal entry. On the other hand, if you charge the black hole, so that there's also electromagnetic contributions to the stress-energy tensor source to the gravitational field, or you rotate the black hole so that it has angular momentum, then the event horizon(s) take on a more complicated form, and the equivalent of the Schwarzschild Kruskel transformations do admit connections. Some researchers have conjectured that this might allow people to move to some other time or place in the universe as through a wormhole, or the connections might just go "elsewhere." Part of the problem is that nobody would live long enough to see you make the trip (see below). The meaning of these solutions can be examined by looking at the trajectory of a geodesic (path of a test particle) falling into the black hole. Inside the black hole, a clock would become connected with the motion along the geodesic to the center of the black hole. It isn't clear whether time would have the same meaning or sensation at that point; nor could anybody ever tell us. An observer outside the black hole would never see anybody fall in. An observer would see someone approaching the event horizon and start falling in. The falling slows exponentially as the person gets closer and closer to the horizon. In conclusion, physics has precious little to say about time travel, primarily because there have been no observed phenomena whose behavior demands the necessity of time travel. Because of this, we don't have a way to assemble machines to allow time travel. Perhaps we can find an already-existing harnessable phenomenon that would allow for time travel.

Appendix 2: Smorgasboard for Computer Junkies

The following simple C-language and BASIC programs are meant to encourage reader involvement and should easily be converted to the computer language of your choice.

Code 1. Lorentz Contraction

See Chapter 2 for further explanation of this code.

```
/*   Experiment with the Lorentz contraction.        */
/*   The code continues to add "9s", i. e. the       */
/*   ratio of velocity to speed of light goes as     */
/*   0.9, 0.99, 0.999, 0.9999 and so on....          */
#include <math.h>
#include <stdio.h>
main()
{
     int           i,   /* counter for each contraction */
                   n;   /* how many contractions */
     double        x,   /* multiplicative factor */
               ratio,   /* statue's velocity divided by speed of light */
              length,   /* initial length of statue of liberty */
           newlength;   /* length of statue after contraction */

     n = 10;            /* compute 10 values */
     length = 151.;     /* length of Statue of Liberty in feet */

     for (x = 1.0, ratio = 0.0, i = 0; i < n; i++, x *= 0.1) {
             ratio += 0.9 * x;
             newlength = length*sqrt(1-ratio*ratio);
          /* for time-slowing factor computation, use: */
          /* newlength =    1.0/(sqrt(1-ratio*ratio));   */
             printf("%f %f\n",ratio, newlength);
     }
}
```

Code 2. High-Speed Rocket for Time Travel

See Chapter 9 for further explanation of this code.

```
#include <math.h>
#include <stdio.h>

main()
{
      double sinh();
      double t, tprime, c, a, year,x;

      c = 2.998e08; /* meters/second */
      a = 9.8;      /* 9.8 meters/second**2 */

     for (year = 0; year <= 20; year++) {
          /* convert years to seconds */
          tprime = (1./4.)*year*3.15e7;
          t = (c/a)*sinh((a/c)*tprime);
          /* convert seconds back to years */
          t = 4*t/3.15e7;
          printf("%f %f\n",year,t);
    }
}
```

Code 3. Gravitational Time Dilation Near Black Hole

See Chapter 10 for further explanation of this code.

```
/* Compute Gravitational Time Dilation */
#include <math.h>
#include <stdio.h>
main()
{
    float ratio, /* circum/circum(hole) */
    time2, /* elapsed time, near hole */
    time1; /* elapsed time, far away from hole */
    int i, j;
    time1 = 1;
    i=0;
    printf("C/Ch  Time 1 (days)   Time2 (days)\n");
    for(j=0;j<=20;j++) {
        ratio = 1+pow(0.5,i);
        i++;
        time2  = time1/sqrt(1.0 - 1./ratio);
        printf("%f %f %f \n",ratio,time1,time2);
    }
}
```

Code 4. Quantum Foam Diagrams

See Chapter 15 for further explanation.

C Program Version

```
/* Create quantum foam diagrams */
#include <stdio.h>
#include <math.h>
main()
{
    float r; /* random number, 0-1 */
    short c[513][513]; /* Arrays for holding 1 and 0 values */
    short ch[513][513];
    int size, time_steps, steps, i,j,k,sum;
    size = 80;  /* Use larger sizes for nicer images */
    /* Controls how many steps foam is to evolve */
    time_steps=20;
    /* Initially seed space with random 0's and 1's */
    for(i=0; i<=size; i++)
      for(j=0; j<=size; j++) {
         r = (float) rand()/32767.;
         if (r >= .5) c[i][j]=0; if (r <= .5) c[i][j]=1;
      }
    /* perform simulation based on twisted majority rules
       to form foamlike objects in 2-D */
    for(steps=1; steps < time_steps; steps++) {
       for(i=1; i<size; i++){
          for(j=1; j<size; j++){
             /* compute sum of neighbor cells */
              sum = c[i+1][j+1] +
                    c[i-1][j-1] +
                    c[i-1][j+1] +
                    c[i+1][j-1] +
                    c[i+1][j] + c[i-1][j] +
                    c[i][j+1] + c[i][j-1] +
                    c[i][j];
                    if (sum == 9) ch[i][j]=1;
                    if (sum == 8) ch[i][j]=1;
                    if (sum == 7) ch[i][j]=1;
                    if (sum == 6) ch[i][j]=1;
             /* Notice "twist" in rules which destabilizes
                blob boundaries. */
             if (sum == 5) ch[i][j]=0;
             if (sum == 4) ch[i][j]=1;
             if (sum == 3) ch[i][j]=0;
             if (sum == 2) ch[i][j]=0;
             if (sum == 1) ch[i][j]=0;
             if (sum == 0) ch[i][j]=0;
          }
       }
       for(i=0; i<=size; i++) for(j=0;j<=size;j++) c[i][j]=ch[i][j];
    }
    /* If you like, you can make a movie of all frames */
    /* to show foam evolving.                           */
    printf("A plot of last frame of simulation\n");
    for(i=0; i<=size; i++){
      for(j=0; j<=size; j++){
        /* Crude attempt to draw blobs using characters */
        /* Better to use hi-res graphics package        */
        if (c[i][j] == 1) printf("*") ;else printf(" ");
      }
      printf("\n");
    }
}
```

BASIC Program Code Version

See Chapter 15 for further explanation.

```
10 REM Create quantum foam diagrams
20 REM R is a random number, 0-1
30 REM C and H - Arrays for holding 1 and 0 values
35 DIM C(81,81)
40 DIM H(81,81)
50 REM Use larger size, S, for nice images
60 S = 80
70 REM T Controls how many steps foam is to evolve
80 T = 20
90 REM Initially seed space with random 0's and 1's
100 FOR I=1 TO S
120      FOR J=1 TO S
130           R=RND
140           IF R >= .5 THEN C(I,J) = 0
150           IF R <= .5 THEN C(I,J) = 1
160      NEXT J
170 NEXT I
175 REM Perform simulation based on twisted majority rules
180 REM to form foamlike objects in 2-D
190 FOR K=1 TO T
200      FOR I=2 TO S-1
210         FOR J=2 TO S-1
220              REM compute sum of neighbor cells
230              A = C(I+1,J+1) + C(I-1,J-1) + C(I-1,J+1)
240              A = A+C(I+1,J-1) + C(I+1,J) + C(I-1,J) + C(I,J+1)
250              A = A+C(I,J-1) + C(I,J)
260              IF A = 9 THEN H(I,J) = 1
270              IF A = 8 THEN H(I,J) = 1
280              IF A = 7 THEN H(I,J) = 1
290              IF A = 6 THEN H(I,J) = 1
295              REM Notice "twist" in rules which destabilizes
267              REM blob boundaries.
300              IF A = 5 THEN H(I,J) = 0
310              IF A = 4 THEN H(I,J) = 1
320              IF A = 3 THEN H(I,J) = 0
330              IF A = 2 THEN H(I,J) = 0
340              IF A = 1 THEN H(I,J) = 0
350              IF A = 0 THEN H(I,J) = 0
360         NEXT J
370     NEXT I
380     REM Swap values in arrays
390     FOR I=1 TO S
400         FOR J=1 TO S
410              C(I,J) = H(I,J)
415         NEXT J
417     NEXT I
420 NEXT K
430 REM If you like, you can make a movie of all frames
440 REM to show foam evolving.
450 PRINT "To plot the foam, place a dot wherever the"
460 PRINT "C array has a value of 1 in it as you scan"
470 PRINT "I and J from 0 to S."
480 END
```

Code 5. Cross-Section of Traversable Worm Hole

See Chapter 15 for further explanation.

BASIC Program Version

```
10 REM Draw Cross-Section of Traversable Worm Hole
20 B0 = 1
30    FOR R=B0 TO 10 STEP .2
40        Z = B0*LOG(R/B0  + SQR((R/B0)*(R/B0) -1))
50        REM  Print data for plotting upper right
60        PRINT R;Z
70    NEXT R
80    FOR R=B0 TO 10 STEP .2
90        Z = -B0*LOG(R/B0  + SQR((R/B0)*(R/B0) -1))
100       REM  Print data for plotting lower right
110       PRINT R;Z
120   NEXT R
130   FOR R=B0 TO 10 STEP .2
140       Z = -B0*LOG(R/B0  + SQR((R/B0)*(R/B0) -1))
150       REM  Print data for plotting lower left
160       PRINT -R;Z
170   NEXT R
180  FOR R=B0 TO 10 STEP .2
190       Z =  B0*LOG(R/B0  + SQR((R/B0)*(R/B0) -1))
200       REM  Print data for plotting upper left
210       PRINT -R;Z
220   NEXT R
230 END
```

C Program Version

See Chapter 15 for further explanation.

```
/* Draw Cross-Section of Traversable Worm Hole */
#include <math.h>
#include <stdio.h>
main()
{
       float b0, r, z;
       b0 = 1;
       for (r= b0; r <=10; r=r+.2){
          z = b0*log(r/b0 + sqrt((r/b0)*(r/b0) -1));
          /* Print data for plotting upper right*/
          printf("%f %f\n",r,z);
       }
       for (r= b0; r <=10; r=r+.2){
           z = -b0*log(r/b0 + sqrt((r/b0)*(r/b0) -1));
           /* Print data for plotting lower right*/
           printf("%f %f\n",r,z);
       }
       for (r= b0; r <=10; r=r+.2){
           /* Print data for plotting lower left */
           z = -b0*log(r/b0 + sqrt((r/b0)*(r/b0) -1));
          printf("%f %f\n",-r,z);
       }
       for (r= b0; r <=10; r=r+.2){
           /* Print data for plotting upper left */
           z = b0*log(r/b0 + sqrt((r/b0)*(r/b0) -1));
          printf("%f %f\n",-r,z);
       }
}
```

Code 6. Time Machine Diagrams

See Chapter 16 for further explanation.

```
#include <math.h>
#include <stdio.h>

int sum, i ,j, rnd, b[101];

main()
{

/* b[i] = 2 means paper in box 2 */
/* b[i] = 3 means paper in box 3 */

/* Initially "place" all 100 papers in box 2 */
for (i=0; i< 100; i++) {
   b[i] = 2;
}

/* Run simulation for 1000 "minutes" */
for (j=0; j< 1000; j++) {
      /* generate random number from 0 to 100 */
      rnd = ((float) rand()/32767.)*100+1;
      /* swap papers between box 2 and 3 */
      if (b[rnd] == 2) b[rnd]=3; else b[rnd]= 2;

      /* Print out number of papers in box 2 */
      sum = 0;
      for (i=0; i< 100; i++) {
            if (b[i]==2) sum = sum + 1;
      }
      printf("%d %d\n",j,sum);
}
}
```

About the Author

Clifford A. Pickover received his Ph.D. from Yale University's Department of Molecular Biophysics and Biochemistry. He graduated first in his class from Franklin and Marshall College, after completing the four-year undergraduate program in three years. He is author of the popular books *Strange Brains and Genius* (Plenum, 1998), *The Alien IQ Test* (Basic Books, 1997), *The Loom of God* (Plenum, 1997), *Black Holes: A Traveler's Guide* (Wiley, 1996), and *Keys to Infinity* (Wiley, 1995). He is also author of numerous other highly acclaimed books including *Chaos in Wonderland: Visual Adventures in a Fractal World* (1994), *Mazes for the Mind: Computers and the Unexpected* (1992), *Computers and the Imagination* (1991), and *Computers, Pattern, Chaos, and Beauty* (1990), all published by St. Martin's Press—as well as the author of over 200 articles concerning topics in science, art, and mathematics. He is also coauthor, with Piers Anthony, of the science-fiction novel *Spider Legs.*

Pickover is currently an associate editor for the scientific journals *Computers and Graphics, Computers in Physics,* and *Theta Mathematics Journal,* and is an editorial board member for *Speculations in Science and Technology, Idealistic Studies, Leonardo,* and *YLEM.* He has been a guest editor for several scientific journals. Editor of *The Pattern Book: Fractals, Art, and Nature* (World Scientific, 1995), *Visions of the Future: Art, Technology, and Computing in the Next Century* (St. Martin's Press, 1993), *Future Health* (St. Martin's Press, 1995), *Fractal Horizons* (St. Martin's Press, 1996), and *Visualizing Biological Information* (World Scientific, 1995), and coeditor of the books *Spiral Symmetry* (World Scientific, 1992) and *Frontiers in Scientific Visualization* (Wiley, 1994), Dr. Pickover's primary interest is finding new ways to continually expand creativity by melding art, science, mathematics, and other seemingly disparate areas of human endeavor.

The *Los Angeles Times* recently proclaimed, "Pickover has published nearly a book a year in which he stretches the limits of computers, art and thought." Pickover received first prize in the Institute of Physics' "Beauty of Physics Photographic Competition." His computer graphics have been featured on the cover of many popular magazines, and his research has recently received considerable attention by the press — including *CNN*'s "Science and Technology Week," *The Discovery Channel, Science News,* the *Washington Post, Wired,* and the *Christian Science Monitor*—and also in international exhibitions and museums. *OMNI* magazine recently described him as "Van Leeuwenhoek's twentieth century equivalent." *Scientific American* several times featured his graphic work, calling it "strange and beautiful, stunningly realistic." Pickover holds patents for a 3-D computer mouse and for strange computer icons.

Dr. Pickover is currently a Research Staff Member at the IBM T. J. Watson Research Center, where he has received fourteen invention achievement awards, three research division awards, and four external honor awards. Pickover is also a novelist and lead columnist for the brain-boggler column in *Discover* magazine.

Dr. Pickover's hobbies include the practice of Ch'ang-Shih Tai-Chi Ch'uan (a form of martial arts) and Shaolin Kung Fu, raising golden and green severums (large tropical fish found in the central Amazon basin), and piano playing (mostly jazz). He is also a member of the SETI League, a worldwide group of radioastronomers and signal processing enthusiasts who systematically and scientifically search the heavens to detect evidence of intelligent, extraterrestrial life.

He can be reached at P.O. Box 549, Millwood, New York 10546-0549 USA. Visit his web site, which has received over 150,000 visits: http://sprott.physics.wisc.edu/pickover/home.htm.

Index